Self-Creating Language

Aleksander Marcisz

TRAFFORD

© Copyright 2002 Aleksander Marcisz. All rights reserved.

No part of this publication may be reproduced, stored in a retrieval system, or transmitted, in any form or by any means, electronic, mechanical, photocopying, recording, or otherwise, without the written prior permission of the author.

For information, contact thebook@secral.info

Some additional information on the subjects discussed in the book can be found on the Internet. Please visit www.secral.info

Printed in Victoria, Canada

Cover design by Aleksander Marcisz

National Library of Canada
Cataloguing in Publication Data

```
Marcisz, Aleksander
  Self-creating language
  Includes index.
  ISBN 1-55369-248-9
  1. Consciousness.  I. Title.
  BF311.M378 2002            126         C2002-900762-3
```

TRAFFORD

This book was published *on-demand* in cooperation with Trafford Publishing.
On-demand publishing is a unique process and service of making a book available for retail sale to the public taking advantage of on-demand manufacturing and Internet marketing.
On-demand publishing includes promotions, retail sales, manufacturing, order fulfilment, accounting and collecting royalties on behalf of the author.

Suite 6E, 2333 Government St., Victoria, B.C. V8T 4P4, CANADA
Phone   250-383-6864   Toll-free   1-888-232-4444 (Canada & US)
Fax   250-383-6804   E-mail   sales@trafford.com
Web site   www.trafford.com   TRAFFORD PUBLISHING IS A DIVISION OF TRAFFORD HOLDINGS LTD.
Trafford Catalogue #02-0061   www.trafford.com/robots/02-0061.html

10   9   8   7   6   5   4   3   2

I ask the reader not to agree with my ideas,
But to be without them for a moment.
The moment of their free choice.

Contents

Introduction 8
How to read the book 8
LABORATORY 9
Experiment 10
Laws by which collections are governed 15
Way of thinking, logic 17
Interpretation 19
Levels of interpretation 19
Multiple interpretations 21
Collection 21
Identifier, information transfer 22
Niches 23
Countability, quantity, individuality 25
Similarity, identity 26
Basin of collection 29
Supplying collection, force of collection, information concentration 30
Reinforcing collection, limitations 30
Sculpture 31
Individuality and sculpture 33
Complexity of collections 35
Discrimination 36
Time sequence, potentials area, time 36
Timelessness 40
Point of time, the present, situation 41
Area of concentration 42
Thread of time, key effect 42
Repeatability 43
The beginning and the end 45
Next stage: the end of existence of our system? 46
Area B and Area K 49
Memory 51
Concept, word, symbol 52

Reality and its description 54
Connection between collections 56
Theorem, truth, untruth 57
Cause and effect 59
Cause of action 61
Probability, fortuity 63
Information 64
Speed, propagation of information 65
Entropy 66
Consciousness of individual existence 67
Conclusions from the experiment 67

PRACTICE 69

Interpretation 69
Levels of interpretation in practice 70
Parallelism of interpretations 71
Canvas 71
Space and time 76
Energy, the principle of conservation 78
Travels in time 79
Laws of nature, physical and mathematical constants, logic 83
Particle physics as niche 87
Behaviour of elementary particles 92
Matter 93
Mathematics 95
Universe 96
Man 98
The beginning and the end with reference to man 103
Society 107
Entropy and man 108
Artificial intelligence? 109
Human niches 110
The real, reality, man's surroundings 112
Imagination, thought 114
Desires, emotions 116
Morality, the good 125
Responsibility 126

Conscious formation of threads of time 130
Habits 131
Contacts of man with other collections 132
Scale of information concentration 134
Supercollections supplied by people 135
Supercollections supplied by other collections than human 137
Free will, active action 138
Destiny? 140
Example with a man and a vase 143
Example with elementary particles 146
Example with an electron, a proton and an atom 150
Example with society 152
Comparison of the Secral with computer science programming languages 158
Knowledge 161
Ways of acquiring knowledge, systems of knowledge .. 163
How should the Secral language be understood? 165
What results from the presented description? 167

CONCLUSION 170
More important concepts used in the book 172
Synonyms of concepts 174
Index 175

Introduction

In the book the reader will find a description of a language. I have not invented the language and no one else has done so. The way, in which it has originated reminds one of how from a little seed there arises something that nobody would expect it to develop.

My role has been to observe and record its successive stages of development. I have also inserted my comments about and interpretations of the observed phenomena.

How to read the book

The reader will find two main parts in the book. The first, entitled 'Laboratory', describes the structure of the language from a theoretical aspect, the second, entitled 'Practice', deals with more practical aspects. Their subsections are often similarly titled since they deal with the same things but described from a different point of view. The language presented in the book does not precisely separate the theory from practice since this results from its nature. Therefore the above-mentioned division of subject matters is not accurate either.

Chapters have been arranged in such an order that the next ones result from the previous ones. However, reading them in another order should not create any problems in understanding.

LABORATORY

When beginning to do anything, we should do it in full, we should be aware where to seek for that beginning. The whole idea of our intentions is contained in the previous statement. The goal of our work will be the creation of something from nothing. If it is not we who create, at least we will observe this process.

Let us not work in accidental surroundings. Let us create a suitable environment – a laboratory in which we will perform an experiment.

Principle 1

'Entities should not be multiplied unnecessarily'

William de Ockham

I have adopted these words as <u>THE ONLY</u> principle of conduct. It is so important that I should like to write it before every single sentence of this book.
As its complement, I will add:

Principle 2

If you have already created an entity, do not simultaneously create any limits to it. Limits are also entities.

A glance at our laboratory will allow us to understand why these principles are so important. I should like us to begin with zero, with nothing. Not even with zero for zero is already a concept and in the beginning we do not want to have ANY CONCEPTS such as 'nothing' or 'nonentity'.

Experiment

Beginning our experiment of creation, let us reject concepts familiar to us and let us forget about everything that we have learned until now. Let us remain only our organ of sight and the most necessary fragments of our reason in order that we can merely observe and remember.

Stage 1

Our laboratory is less than what we commonly understand by the word 'blank'. This is information blank. This denotes not only the absence of all objects but also concepts, ideas and any information. Let us imagine this, let us look at it. What can we see?

Nothing. Nothing happens.

To be precise, one should notice that the concept 'nothing' does not exist in our laboratory either. Likewise the very concept of existence itself and nonexistence.

In order to describe this state one should need any language of description and such one does not exist here.

Stage 2

There is no need for waiting any longer, let us introduce into this state something that is the simplest – an object without any attributes, without any structure. Let us introduce the least possible unit of information. Let us not identify it with a bit since, at this moment, we do not know this concept yet.

Still nothing happens. As observers, we can only ascertain that the object exists. Our system is static.

Stage 3

We introduce another object.

Instinctively we expect that the objects will react somehow to each other. Let us see if this is true. In order to contact one object with another one, the former must have some information about the latter since, as we remember, with the exception of the two objects, there is NOTHING in our laboratory for the time being.

In other words, in spite of the fact that two objects already exist still nothing happens since none of them has any information about the existence of the other. One object cannot transfer the information about its existence to the other one since it would have to possess some information about the latter. <u>The information must be in the object already during its formation.</u>

Stage 4

In that case we furnish one of the objects with some information about the other one. The supplied information is evidently also an object of the same kind as the two already mentioned. The information will play the part of an identifier of Object-2 known to Object-1.

At the present moment we have three objects in the laboratory: Object-1, Object-2 and the identifier of Object-2.

The identifier of Object-2 belongs to Object-1 (enters into its composition). At present, Object-1 is already able to send information to Object-2 since it possesses its identifier. But what information it could send if the only information it has at its disposal is the identifier of Object-2? The situation remains static.

In order to change the static status, Object-1 must dispose of some information which it could send to Object-2. The object itself is not able to produce any information (in reference to what?). The only information, which it could send to Object-2, is that of a third object or its own identifier.

Stage 5

We are adding the next object. Let it be the identifier of Object-1 contained in it. Now Object-1 can send some information to Object-2 in the form of its own identifier. It has done so. At the same time we have observed the first action in our laboratory. At present we can already consider the identity of that object. Looking at Object-1 we can ascertain that it is a creation consisting of other objects.

Object-1, though composed of two identifiers, constitutes a system of three objects since we treat it itself as an object different from its components.

Such approach to the object seems to be different than the one we know from common situations. After all, let us remind that in the beginning of our experiment we have promised not to refer to it. Furthermore, exactly in this manner we create objects familiar to us from our everyday life. Let us, for instance, look at a pencil. It consists of a core and casing. Treated quite separately, they make two objects. Treated as a pencil, they are one object. If, however, we analyse its structure, we will find not two but three objects. The third one, often ignored, is the factor that causes us to treat two objects as one. This factor is identity established by the observer whose subject are two objects, in this case a core and casing.

Object-1 represents the minimal object, which we can observe in our laboratory. We can establish that every object is a system composed of at least three other objects.

Identifiers forming Object-1 are also objects consisting of other identifiers. Object-2 should be similar. Discrimination between objects and identifiers is relative since I distinguish them merely for the convenience of the description. Every object can be treated as an identifier of another one.

In this manner, an amount of information, which is difficult to be analysed thoroughly and ultimately, has appeared in our laboratory. An opportunity has emerged to exchange information in an avalanche way where every individual piece of information is a composition of others previously transferred. This is sufficient to call the system dynamic and any more additions are unnecessary.

Conclusion:
The least possible system of a dynamic character is a system consisting of at least three objects. Each of these objects is a collection of three other ones at the very least.

Henceforth, let us adopt the name *collection* instead of object since it better renders the nature of this phenomenon.

This seemingly leads to a paradox – collections are nested in themselves infinite number of times. This problem arises only when we consider the whole situation from the external observer's view. Let us remember the first principle and let us not create external observers. From the point of view of an object, its whole surroundings are composed of the finite number of identifiers. Thus, from this viewpoint the problem of infinity is merely a theoretical one, created by us as independent observers.

A subjective point of view of an individual object results from the limited information to which it is subject. The object does not penetrate into the internal structure of every subsequent one but it comes to a stop at a certain place of this analysis. The existence of this kind of information negligence is both an attribute and condition necessary for the information interchange between collections. Thus, we can determine that a collection is local negligence of the information. This way of negligence shall be referred to as *interpretation.* Interpretation is an observation characterized by a loss of a certain quantity of information. As a result of interpretation an image is formed, different than that which would result in the case when complete information about the observed object was possessed.

   Let us notice that our system is self-creating and, as a matter of fact, we did not have to create objects since each of them exists thanks to identifiers possessed by other objects. The information interchange between collections and the interpretation of many collections as one, result in separating new collections. Such an uncommon exchange of operations can occur as long as preferred.
   The previous conclusion, stating that the minimal dynamic system is a system consisting of three objects is also reflected in the idea of transferring information, i.e., identifiers. The concept of transfer assumes the existence of a sender, a receiver and a transferred object.

   What is the identifier of a given collection? It can merely consist of the information, which a collection possesses. Therefore, <u>the collection will build its identifier from identifiers of other collections known to it</u>. Its identifier determines what other collections are known to the given collection and this fact distinguishes it from others.

   Let us notice that a collection with several identifiers can gather them in any combination, forming simultaneously new objects. What is such a new object? This is also an identifier since it is the only kind of object, which is known to us in our system. A newly formed identifier represents a definite object yet unknown in our system. The identifier may, but it does not have to be sent to the other external collections. What may the further fate of the new object be? If its parent collection sends it some other identifiers known to it, the new object will become more complex internally. The more complex the new collection is, the more identifiers will be sent to it and also the information interchange within it (between its components) will increase.
   How does the internal object differ from the external one? No imposed discrimination exists in our system, which would allow

confirming that. The only discriminant is the information interchanged with other collections.

A collection is interpreted as an internal one when its identifier enters into the composition of the identifier of the parent collection and when the information interchange with the parent collection is greater than the total information interchange with other collections.

A collection is interpreted as an external one when its identifier does not enter into the composition of the identifier of the parent collection, then from the viewpoint of the parent collection, the collection does not exist. This is also the case when information interchange with the parent collection is less than the total information interchange with other collections.

An identifier is not unique, since it is built from these identifiers which the given collection possesses and it does not ensure the unique character of the identifier. However, this does not seem to be any obstacle for the existence of our system.

Laws by which collections are governed

An object, which we observe in our laboratory is a collection of information or, in general, it is information. Information exists (it is readable and it is possible to confirm its existence) only when it is being <u>transferred</u>.

Thus, it is possible to confirm the existence of a collection only when it is transferring information to another one.

Conclusion: One is not able to confirm the existence of an individual collection without the existence of concurrent collections and this is the equivalent of the non-existence of the collection.

Information does not originate from anything and it cannot be created. The only thing that can be done with information is to transfer it, i.e., to send or receive it.

Beginning the observation of our laboratory situation, we have displayed a certain ingenious opinion by stating: "at present we are adding the next object". As observers, we should not have done it. I let myself do it since this description seemed to be visual for me. We have got rid of the foregoing naivety after having stated that the collection does not exist 'itself from itself' but it results from the existence of the other ones. A problem of the kind: "where did the first collection arise from?" does not exist in our laboratory for the time being, since by asking such a question one assumes the existence of time. Such a concept is yet unknown in our laboratory (let us remember Principle 1).

Some naivety is still left in our prior reasoning. This concerns 'information transfer', i.e. the exchange of identifiers. The action of transfer is something completely new in our laboratory and we are determined not to create rashly new entities (Principle 1).

The next naivety regards our role as observers. However, if we have already a general concept about the situation in our laboratory, we can abandon this inconvenience. It is enough if we place ourselves within this situation. That is, from this moment an observer will be one of the collections, which has been dealt with before. From that moment on, we are already not in a privileged position in our laboratory. We are subject to the same laws as every collection. Owing to this, we have liquidated the next entity that we created before, that is an independent observer.

Has the above-mentioned naivety been liquidated as well? Not entirely. Though we ceased to exist as independent observers, we have retained our way of thinking. From this moment we should also treat the way of our reasoning as one of the elements of the situation. Let us remember this since even if we reconcile ourselves to the fact that the independent observer cannot exist, we treat our ideas, concepts and all our way of thinking as something independent of the other phenomena.

Logic is the way of thinking which we acknowledge most readily. Let us look at it closer.

Way of thinking, logic

The way of thinking which we use is founded on identity (*this* is *this*...).

The whole of logic is merely based on this one operation. If we separate some other operations, we will notice that they are more or less sophisticated compositions of individual operations of identity.

In every law and theorem we will find the word 'is'. In every mathematical formula we will find an equality sign or its equivalent. Every sentence that we say, write or think contains the word 'is'. If it does not occur directly, it occurs conjecturally as the ascertainment of existence. Identity is never a single-argument operation. For instance, the statement "A exists" seemingly refers to one object 'A'. After all, let us not forget the definite observer. Let us notice that the statement deals with three objects: an observer, the object named 'A' and the concept 'exists'. In its complete form, the statement should state "A exists for the observer B".

An operation contrary to identity does not exist. The statement "A is not B" emphasizes differences between the two objects but it still constitutes their identification. We will deal with this phenomenon in a more detailed way in further chapters of the book.

Identity is an inseparable attribute of the information transfer. <u>The information transfer is the act of identity performed by a sender. The object of identification is a receiver and the sent information.</u>

<u>One can find the existence of information only by transferring it</u>. The existing information is solely that which is being transferred.

As a conclusion we may write down:

Element of information = Act of identity

The operation of identification is a manifestation of consciousness.
According to what we have stated above on information and identity, we can confirm that <u>the act of identity is the elementary process of the phenomenon of consciousness</u>. This is an elementary act of thinking.

Process of consciousness = Process of identification

An individual act of thinking is an act of identity and an element of information.
This may be combined in one formula:

Element of information = Act of identification = Act of consciousness

Since the element of information is nothing else but a collection, which we described in our laboratory, we can insert still one more member into the above formula. ***The collection*** represents the same thing as the previous elements of the equation.

According to the above remarks and our present knowledge about collections, we can state that <u>consciousness is a fundamental phenomenon for the behaviour of collections and their inseparable attribute</u>.
Acknowledgment of consciousness and thinking as biological or psychological phenomena is a secondary interpretation.

By describing the way of our own reasoning and treating it as the element of a situation, as one of collections, we have disposed of the naivety mentioned in the previous chapter and which made an external observer from our way of thinking.

We have also got rid of the necessity for treating the actions of the information transfer as something fundamentally different from the information itself and from a collection.

Interpretation

As we have noticed earlier, *the interpretation* is an observation connected with the loss of a portion of information. As a result of the interpretation the observer has a different picture than that resulting from the full information about the observed object.

In other words, the interpretation is building a new concept on the grounds of those already known.

By building a new concept we ascend a high level of interpretation.

Levels of interpretation

They reflect our way of treating collections, i.e., they determine the quantity of information lost during observation.

As the first level of interpretation we will assume treating collections as elementary operations if identity without any supplementary interpretations and without any loss of information.

Interpreting collections as something different, we ascend the next, higher level of interpretation. Some collections can adopt the role of attributes characterizing other collections. A further development of interpretation by integrating many attributes into one is a further level of interpretation.

It is irrelevant to define here precisely each of the levels but one should be aware of the fact that the ways of treating the same collection can vary. Treating white colour, as the feature of an object as we do it every day, is a higher level of interpretation than treating the colour as a collection devoid of the attributes and which merely interacts with other collections. Higher levels of interpretation make detecting collections and the nature of their action difficult whereas they make easier what we refer to as 'everyday life'.

It is particularly important for us to understand the levels of interpretation when attempting to describe any observations. Each of our descriptions is composed of concepts and if a concept is to serve us well, we should understand well what the concept itself and descriptions are, as well as by what laws they are governed. The concept and description are collections like the described object itself. In order to describe something in a correct way one should begin with the lowest level of interpretation since the number of concepts occurring on this level is small and allows for defining them accurately. Developing these concepts in a controlled way, we can ascend the further levels of interpretation (closer to our everyday ideas), preserving a relatively compact, well-defined structure of the whole description.

On the other hand, if we made an attempt at producing a description beginning with the concepts of high levels of interpretations, we would create a structure founded on obscurely defined concepts. The final effect would at least be as faintly and abstrusely defined as its constructive element. For instance, we often employ such concepts as 'quantity', 'truth', 'obviousness' but no explicit, commonly accepted definition of these concepts exist. They are poorly defined and differently understood by various people. If we try to build a description of something, on the basis of such concepts, we will not provide description more accurate than these concepts. Therefore we have begun this book with an experiment of forgetfulness of all concepts and their gradual development.

Multiple interpretations

Overlapping the successive interpretations is always connected with the increase in the loss of information available in our system. This also happens in case of interpretations treated as their contradictions.

For example, the application of interpretation connecting two collections into one and then splitting such a formed collection again in two is not equivalent to the return to the original situation. The collections, obtained in this manner are not the same as those connected with the first interpretation on account of the loss of information in every subsequent interpretation. Their return to the original situation is possible merely by abandoning the interpretation but not by forming new collections.

Collection

This is a concept according to which all the others are developed. It cannot be defined by use of other concepts. All of them are secondary in relation to the concept of collection. In simple terms, collection can be determined as an elementary stuff of everything that is observable.

A collection is not describable by means of other concepts but we, as people need such a description, which we could exchange. Thus, we will make attempts to describe a collection.

A collection is a starting-point for developing other concepts, descriptions, phenomena, objects, etc. This will permit us to be consistent with using concepts and making descriptions. The lack of consistency would result in producing descriptions as poorly defined as their foundations. <u>Precisely this and only this is the</u>

reason why I have used Principle 1 at the beginning of the book. Principle 1 is not any external law, which we impose on ourselves. In reality one should not even mention it, since it simply illustrates the consequence of developing concepts.

The present book almost as a whole is a definition of the concepts of a collection.

Identifier, information transfer

An identifier is *a collection*, which is subject to the same laws as the collection, which is represented by it.

The identifier as a collection is not an invariable object but it undergoes some changes. Therefore, it is not necessarily a unique identifier.

In order for the object to be able to send information to another one it must have its identifier. Consequently, it may not acquire an identifier immediately from the object of destination. It can attain the identifier solely from a third object. After obtaining the identifier the object can already be developed without mediation of other objects.

Information transfer to a collection consists in sending information to its identifier. In other words, this relies on identifying the identifier with the collection, which represents the information being sent.

Let us remember that a collection does not exist of itself but its existence is based on the fact that other collections possess its identifier. And merely owing to this fact, it is possible to find the existence of a collection and any interaction with it. Therefore, any alteration to an identifier denotes a simultaneous change of the collection itself represented by the identifier. The concept of 'information is being transferred to a collection' reflect this property of identifiers. The concept of 'information transfer' does not very well describe this process but we can abide by it since no other term will fully render the occurring process and it will allow us to understand it.

We can call an identifier strong or weak depending on the number and complexity of collections from which it is composed. A number of collections supplying the identifier (performing the acts of identity with the identifier taking part) determine whether the identifier is strong. The stronger the identifier is, the stronger the collection represented by it. It happens because the existence of a collection consists in the existence of its identifiers.

The stronger the collection is, the more difficult it is to be altered. In order to do so a relatively great amount of information should be sent, i.e. an altered identifier should be sent to many other collections. This may be a difficult task for an individual collection. The success depends on the number of identification operations it is able to perform at a given time.

On the other hand, since the existence of collections consists in performing continual operations on identifiers, <u>every identifier in conjunction with the collection that it represents changes permanently</u>. Continual alteration results from the fact that in order to make observation of a collection possible, the operation of identification must be performed on its identifier and this denotes the alteration of an identifier and the collection, which it represents. This alteration will be insignificant in case of strong collections but it will occur in every case.

Niches

A collection is able to exchange information with the collections which identifiers it has.

Not with each of the collections the information interchange is equally intense by reason, among others, of the differences in the degree of the development of particular collections. One can separate a group of collections with which the information interchange will be more intense than with the others.

These facts lead to the concept of **an *information niche***, i.e., the area of operation of a given collection. This area is made up of a group of collections with which the given collection exchanges information more intensely in comparison to the other ones.

Collections occurring beyond a niche either do not participate in the information interchange or this interchange is feeble.

The reasons why a collection is beyond a niche may be the following:
1) The identifier of a collection, possessed by members of a niche is weak. The information interchange is of low intensity. A collection is interpreting the situation, i.e., the observer decides which intensity is weak and which is strong.
2) A lack of an identifier.
   The immediate exchange is not possible until the identifier is acquired through the mediation of other collections.

In the first case, marking out the limit of a niche is conventional, dependent on what we understand by a 'feeble identifier'.

In the second case, the limit of a niche is more distinct although here one may talk about the 'force' of the limit. This is affected by the difficulty in acquiring an identifier of the collection occurring beyond the niche (the difficulty, i.e., through mediation of how many other collections one can attain the identifier).

Every collection occurs in at least one information niche. Every collection constitutes a niche for its component collections.

A collection may belong to many different niches, which not necessarily must be enclosed one in another. However, by analysing information interchange of a definite collection with the other ones and comprising the larger and larger areas of collections grouped in niches, we will arrive at the conclusion that for every collection a *'hierarchy of niches'* can be determined, in which this one is contained.

The nearest niche of a collection is the collection itself. This niche is formed by collections entering into its composition. A distinction between the concepts of a collection and a niche results from interpretation. A collection is something fundamental belonging to the first level of interpretation. A niche is a remarkable interpretation of a collection and it belongs to a higher level of interpretation than the first one.

Countability, quantity, individuality

A quantity is not an original concept. It has not arisen on the first level of interpretation. I have not referred to this matter previously, in order not to make understanding of what we observe in the laboratory difficult. A collection in itself does not have such a feature as quantity and it cannot be stated that a given collection is 'one' or attribute any other quantity to it.

From the beginning of our observation we have assumed that collections are distinctly separated and discriminated objects from each other. We have assumed such an introductory way of treating a collection for the convenience of our description. Now we must loosen a bit this strict treatment and, as a matter of fact, we must be more precise about our concept of a collection. Another way of interpretation is understanding collections as unstable 'objects', not separated distinctly from each other and impossible to be discriminated fully (impossible to be counted).

None of these interpretations is better or worse. They can be used interchangeably and linked together depending on the situation. Here the only criterion is to understand the description produced by us. Whichever interpretation is applied, we must not forget about the existence of another one in order that our descriptions are not too vague (we remember Principle 2).

We have qualified the concepts of continuity and separation as those possessing the same descriptive values, since they occur on the same level of interpretation, as their reciprocal contradictions. Desiring to provide a better description, the observer should descend to a lower level of interpretation, rejecting both these concepts.

The concept of quantity arises as a result of neglecting the weak correlations between collections. Neglecting the feeble connections enables the limits between collections to separate and they can be treated as countable objects, as individualities.

The interpretation of countability leads to arising the concept of number and, consequently to arithmetic.

The interpretation of space, in general understanding, arises when whichever characteristic is interpreted as the one adopting the values of continuum. Then we talk about the 'space' of this feature. Its space appears together with arising the interpretation of countability. A quantity is a feature adopting different values and forming the space. This is the space, which is an archetype of spaces familiar to us and also to our interpretation of spaces, which we call physical or geometrical. One of its properties is the quantity of dimensions. Talking about the quantity of spatial dimensions is a simplified interpretation of the continuity of this feature arisen when countability is applied to dimensions.

Similarity, identity

Two collections are identical when a collection, playing the role of the observer, is not able to find any distinctions between them. The identity is relative and depends on the observer to a large degree. If collections become identical, they begin to be interpreted as one collection. This happens so, since the quantitative discrimination of collections is founded on differences between them.

The same thing, which we have stated about identity, refers also to the concept of similarity of collections. Identity is an extreme case of similarity.

The acknowledgment of two collections as identical or comparatively similar denotes that an identifier sent to one of them can reach another one. The more the collections are interpreted as similar to each other, the more distinct the described effect will be. Only in case when the identifier of a collection is explicitly different from the others, one can talk about a reasonable certainty that every information sent to the collection indicated by the identifier, will really reach it and not any other regarded as similar.

In case of the intense information interchange with a definite collection, in practice we will observe a stream of information spreading on many similar collections. The definite collection, as the actual goal for the information interchange, will be that one to which the information is directed most often, i.e. it will be the main running stream of information. In case of very similar collections it may be difficult to find which one is the main current.

Let us remember the effect described above as **the principle of ambiguity of identifiers**. In short, we can write it as follows:

An identifier represents one or more collections to a degree corresponding to the similarity of the given collection to the identifier.

As a conclusion from this principle let us write:
<u>Every collection, similar to another one, can play the role of its identifier to a degree determined by their reciprocal similarity.</u>

As the description of similarity, let us assume:
Similarity is defined by the reverse of the quantity of unique information possible to be transferred between collections.

This denotes that the collections, which are in a position to exchange only a comparatively small quantity of unique information, are treated as similar to each other.

Let us remember that the information transfer to collections is not similar to the transfer of physical objects. The information sent to another collection is still in possession of the sending collection. <u>Sending the same information to the very same collection for the second time will not altogether be interpreted as the information transfer, since no alteration in the collections will be registered</u>. The more the two collections differ from each other and the greater concentration of their information is, the more intense information interchange can occur between them. And reciprocally, the more similar collections are to each other, the weaker information interchange between them will be. Intense interchange leads to the reciprocal assimilation of collections. As the information is transferred between two collections, the intensity of interchange will weaken because of the exhaustion of the unique information possible to be transferred.

We can talk about the definite ***information potential*** of a collection with regard to another one. This potential defines the quantity of the unique information that may be transferred between collections.

In other words, the information potential determines a distinction between collections.

The described effect which makes sending the same information to the very same collection impossible for the second time is not the 'interdiction' or principle coming from the above but it only results from the interpretation of both sent collections as identical.

In the chapter 'Way of thinking, logic' we have stated that an operation reverse to identity does not exist and that the statement "A is not B" merely stresses the differences between objects constituting their identification.

Now we can explain it more accurately. The statements "A is B" and "A is not B" register the relation of similarity between two objects. The act of comparison of two objects is sending identifiers of a collection to them, which makes the comparison. This is merely enough to find that these objects are not perfectly different, since some of their components are similar. The formula "A is not B" consequently ascertains great differences between the objects and, after all, is the very same operation of identity as "A is B" and not its contradiction. It becomes one only after applying the interpretation of this kind as used in relation to every day statements of that kind.

Basin of collection

A definite collection exists since other collections have its identifier. A set of such collections will be called the *'basin'* of a definite collection, analogically to a river which existence is founded on its basin. As not seldom we have a problem to determine which of the arms of a river basin is the headspring, also in case of a collection it may be difficult to establish the same thing.

This concept has been introduced because it draws our attention to the characteristic feature of a collection and it will be helpful in our further considerations.

The larger the basin of a collection is, the more information must be transferred in order to alter a given collection in an observable way. Remembering that individualization of a collection is merely one of interpretations derived from a higher level than the first one, let us notice that the basin cannot be defined with a greater accuracy than the collection itself. Marking out the limits of a basin depends on the way we use the interpretation of similarity since the basin is defined by identifiers similar to each other.

The basin does not have to be restricted to a niche in which the collection occurs at present.

We will often employ the concept an *'extended identifier of a collection'* or an *'extended collection'*. By **extended** I mean *possessing an extended basin.*

Supplying collection, force of collection, information concentration

By a ***strong*** or ***weak*** collection I understand the collection that has a comparatively well-extended or unextended basin.

By ***supplying a collection*** I understand such an act of identity in the result of which the basin of the collection will be extended.

The force of a collection, ***the information concentration*** – I use these two concepts interchangeably in order to present figuratively the size of its basin.

The extended basin, i.e. a strong collection needs sending a large quantity of information in order to modify its attributes in a visible way.

Reinforcing collection, limitations

Sending information to an identifier of a given collection extends it in the sense of information. In other words, it reinforces the identifier. This may only be the unique information such as one that the identifier has not had yet. From the viewpoint of the supplying collection, this limits reinforcing the collection. The given collection can be reinforced inasmuch as it has the unique information possible to send.

If the reinforcing collection does not limit itself to mere

reinforcement but also controls its effects, it receives information from the collection being reinforced. The assimilation can be neglected if the quantity of sent information is insignificant as compared to the whole of the information forming the collection.

A reverse operation to the act of identity does not exist therefore there is no operation attenuating the collection. The attenuation decreases the quantity of acts of identity with a given collection, i.e. it lessens the basin of the collection. Since the basin is formed by more than one collection, a great number of collections will have to be involved in the process of attenuating the definite collection.

Sculpture

Let us have a closer look at the formation process of collections. A new collection is formed by connecting others (i.e., by interpreting many collections as one). It has its identifier, which, at the beginning, is only in possession of the forming collection. This identifier can be next sent to other collections. Let us notice, that if a new identifier does not differ much from another one already known in a niche, according to the principle of ambiguity of identifiers (the chapter 'Similarity, identity') a considerable quantity of information being transferred to the new identifier will reach another collection. Every piece of information reaching another collection makes the identifier similar to it. This will assimilate it and gradually replace the new identifier with the already known one.

Eventually, <u>we will observe the difficulty in forming collections different from those already known. However, if such a collection has been successfully formed, in this manner registering itself in the sculpture of the niche, the formation of new, similar collections it will henceforth be facilitated.</u>

Let us remember that collections are not merely static objects, which we observe but they are also the whole situations, i.e. a composition of many collections occurring in a definite time sequence. Differentiation between static objects and situations is merely another view of the same phenomenon, it is another interpretation. Everything that has been described above also concerns those collections that are interpreted as dynamic phenomena and situations.

Therefore, noticing a new phenomenon or a little known one in the niche will be more difficult than noticing a known phenomenon, i.e. such one, whose identifiers are in possession of many collections of the niche. An attempt to induce a new phenomenon will tend to evolve into observation of one of the phenomena already known in the niche. After all, a successful attempt to induce a new phenomenon will facilitate its recurrence provided that an identifier of the phenomenon will be sent to a large number of collections occurring in the niche.

The collections existing in the niche force, in the above-mentioned manner, assimilation of the newly formed collections into those already existing. This effect is exerted by the collections most similar to a newly-formed one to a degree corresponding to the magnitude of the resemblance. The action of these collections resembles the activity of a matrix or a mould used for the newly-formed ones. A group of collections that acts on the definite collection in this manner shall be hereinafter called **the sculpture** of this collection. By **the sculpture** of a collection, one can also understand its shape, which it will acquire if the assimilation into the mentioned group is complete.

The whole set of all such sculptures for all the individual collections occurring in the niche shall be called the sculpture of the niche.

The name *sculpture* renders the dynamic nature of this effect similarly to the sculpture of a river bed or streams forming caves. The sculpture of the niche is subject to continuous variation. The change results from the acts of identity constantly performed within the niche.

Summing up, I use the concept *sculpture* in two meanings:

The sculpture of collection – These are all the collections similar to the given collection and contained in the niche. Because of ambiguity of identifiers, the identifiers from this group account for forming a concrete collection. Expressing it picturesquely, this is as if it is a kind of a collection, e.g. a plum-tree growing outside the window is subject to the influence of the sculpture also determining the features of all the other plum-trees. It is also subject to the influence of a wider sculpture which is the sculpture of a tree and the sculpture of plants, etc. These sculptures have been formed in the present niche in which we find ourselves.

The sculpture of niche – It is a set of all the sculptures of collections (from the previous item) taken altogether and occurring in a given niche.

The way in which we perceive collections is forced by the sculpture developed in our niche. We see a certain collection as a plum-tree. Collections, occurring in other niches, can perceive our plum-tree in a completely different manner dependent on that sculpture.

Individuality and sculpture

As it has been stated in the previous chapter, during the information interchange with an identifier in case of existence of many similar collections one cannot be sure with which individual collection the interchange is held. Such an exchange of information, repeated many times, will not deal with one collection only but the whole group of similar collections. Such a group was named ***the sculpture*** of a collection. The sculpture of the collection, as the average of the attributes of its component collections, is a kind of a pattern or a matrix for each of its representatives. Sculptures of all the collections within the given niche form the sculpture of the niche.

<u>Supplying the given collection we always supply its sculpture</u> since the collection itself enters into the composition of this sculpture. On the other hand, we do not necessarily supply the definite collection since on account of ambiguity of identifiers this action may concern one from other collections of the sculpture.

The sculpture is nothing else but a separate niche for its representatives. The described <u>process of forming the sculpture can also be understood as a description of forming the niches</u>. However, let us remain with two different names 'sculpture' and 'niche' since each of them focuses on another characteristic feature of this phenomenon.

As something opposed to the sculpture, one can recognize a single collection as a representative of the sculpture. A single representative will be called *'individuality'* or *'individual collection'*. Individuality differs from the sculpture and it has also another basin.

A degree of the extension of an individual collection and its sculpture can be different and independent of each other. The strongly extended individuality can have a weakly extended sculpture. This denotes that not many collections similar to it exist. The reverse situation is also possible: a strongly extended sculpture and many weak individualities slightly different from each other.

The continual information interchange only with a certain group of collections can bring about the loss of individuality for the benefit of a sculpture. We will observe reinforcing the individuality if information interchange of a given collection is made with the collections of great reciprocal discrimination.

Complexity of collections

It is possible to talk about complexity when we know the interpretation of countability. The statement: "Collection D enters into the composition of collection C" denotes that information interchange of collection C with collection D is comparatively greater than interchange with other collections with which these collections communicate. The collection, interpreting the situation decides how great this relation is to be. According to the quantity of information about collections C and D possessed by this collection, one of them can be interpreted as a component collection of the other. The very same collection can be interpreted as a component of different collections. The way of the interpretation may be forced by the sculpture of the observer's niche.

A collection is composed of other collections and each of them still exchanges information with the others. All the existing collections are connected with each other directly or indirectly through a chain of the performed acts of identity. Determination, which collections enter into the composition of the given collection, is a relative thing and it is their interpretation.

Component collections exchange information between each other. However, by doing so they have less and less unique information for sending. As they interact, they assimilate and consequently their reciprocal information interchange weakens. So, in order that a collection could preserve its complexity, it is necessary to exchange information between its component collections and the external collections. In this manner, the component collections gain the new, unique information, which they can send among themselves, maintaining information interchange on the level, which permits them to be interpreted as portions of a great collection. In case when this exchange weakens below the mentioned level, the collections will no longer be interpreted as the component ones. On high levels of interpretation this may be regarded as the 'disintegration' of a parent collection, i.e. the end of its existence. On low levels of interpretation we will rather talk about the departure of a given collection beyond the niche in which the observation has been made since it is not possible to ascertain the complete 'disintegration' of the collection. Information interchange between collections can weaken but it will never cease completely.

Being a component collection does not necessarily denote that there is a connection in sense of the near distance in space. This connection may be reflected in the physical space if the interpretation of distance exists in the given niche.

Discrimination

By discrimination I understand the situation in which an individual collection begins to be treated as two or more collections.
<u>The operation of discrimination is the operation of identity with one of the components of a given collection.</u>
Therefore, discrimination is not as fundamental an operation as identity and it is not a contradiction of identity. This concept arose on the high level of interpretation. The identity is the lowest level of interpretation (identity amounts to a collection).

Time sequence, potentials area, time

The interpretation of time arises when different collections are interpreted as one and the very same collection. As a result of such an interpretation, a sequence of collections is formed and interpreted as successive images of one and the same collection but varying in time. This sequence will be called *'time sequence'*.
A collection can be included in a time sequence if its infor-

mation potential with regard to the niche is significant and, at the same time, the identifier of the collection is in possession of the members of the niche.

The sequence can be associated with an individual collection or the niche (let us remember that we can also treat an individual collection as a niche). A collection included in the sequence becomes an object of information interchange with all the members of the niche with which the sequence deals. This denotes that the interchange will be comparatively intense and connected with the decrease in the information potential of the collection included in the sequence with regard to the niche. The decrease in the information potential denotes weakening information interchange and identifying the time sequence with the next collection of high potential. This can be figuratively described as the successive burning of potentials charged with regard to the niche of the collection. The loss of potential with regard to one niche can be associated with its growth in view of other niches. As a result of the inclusion of the collection in the time sequence, its membership of the niche may consequently be altered.

If in a niche the interpretation of time is known, this denotes that the time sequence exists in it. Every collection, which knows the interpretation of time, has its individual time sequence with which it identifies itself.

The very same collection can occur in time sequences of many other collections. By separating a portion common for all individual sequences in a given niche, we will obtain a common time sequence of the niche. The sequence of the niche contains these elements, which occur in every individual sequence of the niche members. The situations included in an individual sequence are observed only by the individual collection, which identifies itself with that sequence.

The situations included in a sequence of the niche are observed by all members of the niche. The condition for including a definite collection into a sequence of the niche is its high information potential with regard to the components of the niche, at least permitting the collection to appear in individual sequences. Therefore its identifier should be in possession of a comparatively large number of members of the niche in the adequately reinforced form as compared to the extension of the whole of the niche.

A collection performs the acts of identity with others and reinforces them to a different degree. In this manner, the area of the information activity of a given collection is formed. One of the collections occurring in this area will be selected as the next element of the time sequence. A group of collections, which can become next elements of the time sequence, will be named the *'potentials area'* of a collection, which is associated with this sequence.

Time sequences constitute the foundation for further interpretations. Time, interpretation of which is formed on the existence of sequences can be 'measured' only as a distinction between the intensity of the successive acts of identity along different sequences. By comparing two time sequences we can find the difference in the frequency of including new collections in the sequence. The difference is a basis for measurement of the lapse of time seen from the viewpoint of one of these sequences. The second sequence constitutes a fiducial point of measurement.

The measurement can be regarded as accurate if a sequence of a greater frequency and stability is adopted as the fiducial point of measurement. Let us notice that if the frequency of a fiducial sequence varies we will not be able to ascertain this fact without a reference to the third sequence. This happens because as a matter of fact we do not measure the frequency of the sequence but the <u>difference</u> of the frequencies of the sequences. We are not in a position to measure the frequency of a single sequence since in this case the fiducial point is lacking. What in everyday apprehension we regard as 'independent running time' is lacking. While measuring time we assume that the sequences, which we regard as a reference, are stable with regard to our own sequence. This assumption can be performed only approximately since every collection exchanges information and it does not remain invariable. The time measurement depends on time sequences, which participate in it and, at the same time, also on the niche in which the measurement is taken. Any measurement is not universal but local and it concerns the definite niche.

It is necessary to stress that time and time sequence are the local interpretations and one cannot talk about any universal time independent of a collection. Every collection is able to control its own or any other time sequence to a lesser or greater degree. By control of a sequence I understand the possibility of reinforcing the selected collections from the potentials area.

When we find ourselves on a lower level of interpretation than our everyday one, time is not anything what 'is running' in the direction from the past to the future. <u>Time is the interpretation of the acts of identity, linking the selected collections dissimilar to each other together, forming time sequences</u>. One cannot talk here about the 'direction of time'. An attempt to 'reverse time', which relies on including the collections which were already in it in the same sequence but in the reverse direction denotes that the existing time sequence goes on in this specific manner. Moreover, when we include a collection in the very same sequence in which it was for some time, we will not obtain exactly the same effect as before, because the identifier of this collection has already been altered. Referring to it for the second time will not necessarily denote the information interchange with the very same collection as beforehand for the sake of the ambiguity of identifiers. In the interpretation of the parent collection this will look as if the collection, included in the sequence for the second time has altered but not itself.

Niches can have a complex hierarchy and their limits may penetrate one another. Their time sequences will also reflect this complexity. The situations observable in a given sequence can be noticed in certain sequences but not in others. This depends upon the connections between niches, i.e. on the intensity of information interchange between them.

<u>Not every niche does have to interpret the very same situation in the same manner</u>. If, in a given time sequence, we observe a situation composed of two collections then in a time sequence of another niche in which the very same situation has been included, we may observe three collections of a completely different sculpture.

The way of interpretation depends on the magnitude of the information being lost and on the formation way of a sculpture of a given niche. If interpretations of the very same situation vary significantly, this signifies a comparatively great difference between sculptures of the niches and, in turn, it denotes a weak interconnection of both niches.

Timelessness

Timelessness, i.e. a lack of the concept of time does not denote duration in a static condition but <u>the freedom to perform the acts of identity</u>. Time denotes the restriction to the acts of identity within the framework of the specific area of a collection called a time sequence.

Thus, the distinction between time and timelessness is not sharp but it may be graded in determining the freedom to perform the acts of identity. In other words, the dependence of a given collection upon the interpretation of time can be small or great. This depends on a degree of freedom, which the collection has in the choice of the goals of its acts of identity.

One can ask the question: How is it possible to reconcile the statement that timelessness is the freedom of choice with the fact that this action itself already assumes that it occurs in some time? When we try to imagine or describe a niche in which the concept of time is not known, we build our idea from the concepts familiar to us, i.e. we create a concept still on a higher level of interpretation than the concept of time. In the meantime we dealt with something contrary, with descending below the level of the interpretation of time. This will not be possible as long as we ourselves as observers are under the influence of this interpretation.

The freedom to identify ourselves brings about abandoning the concept of time which we measure with clocks and our own lives. But this does not mean yet that all the interpretations have been abandoned. Such concepts as 'choice', 'niche' still remain. They allow for the definition of timelessness described above. We are forced to formulate it in the manner suggesting our dependence upon time, since the present description together with the readers and the author is subject to the interpretation of time.

Point of time, the present, situation

Making the measurement of time, a collection compares the frequency of information interchange between collections. In this manner, it is not able to find a shorter period of time than that between successive acts of identity dealing with their own time sequence. The minimal period of time, adequate for a given collection, we will call a *'point of time'*.

The point of time can be interpreted as **'the present'** with reference to the given collection. **'The present'** is a group of collections that has just been included in the time sequence.

We often use the concept of a *'situation'*. As a *'situation'* I call a system of collections, which is observed at a definite point of time by the collection playing the role of the observer. These are collections creating the present. Niches, time sequences and situations are subject to a hierarchy. The situation in a definite niche can be a portion of the situation of a superior niche. A situation itself is a collection. Such collections precisely form a time sequence.

Area of concentration

These collections with which a given collection is exchanging the information most intensely at a given moment can be named the *'area of concentration'*.

If a collection knows the interpretation of the time sequence, the area of concentration will contain all or a part of a group of collections included in the individual sequence, i.e. a group that we have called *'the present'*.

From the viewpoint of a superior niche, the area of concentration of its component collection may also cover a part of the potentials area of a niche. For that reason, the areas of concentration of individual collections are of great importance for the formation of the time sequence of their common niche.

Thread of time, key effect

A collection can be perceived not only as a static object but also as a situation, a composition of many collections in a definite time sequence. The differentiation between static objects and situations is merely another view of the same phenomenon, it is another interpretation.

If a group of collections observed in the definite time sequence is interpreted as one collection, then we deal with the effect, which shall be named a *'thread of time'*. Threads as collections are inclined to group in niches, forming definite **sculptures** within a superior niche as it is described in the chapter 'Sculpture' and 'Individuality and sculpture'. Thus, observing any time sequence we will notice rather recurrent threads of time than collections occurring in a disorderly manner. Though threads of time are observed as recurrent elements of the sequence, their individual component collections can be different every time.

The above-mentioned inclination for grouping and forming a sculpture brings about still another effect. If in the thread of a well-formed sculpture one component element is lacking, we will notice the inclination for forcing the presence of this element. The effect will be noticeable notwithstanding the deprival of its elements. The thread is still comparatively similar to one of the threads contained in the sculpture of a niche. How strong the forced effect will be depends on the degree to which the sculpture of the thread has been formed, i.e. how many acts of identity have supplied it.

Since this effect is so important, we will name it the *'key effect'*. Hence the name appears that a thread of time, deprived of at least one element, can act as a key, forcing the missing element to appear in the time sequence. As *'a key'* or *'a time key'* we will call a thread different from one of the threads known in the niche and varying in a lack of at least one collection but characterized by a remarkable resemblance to the known thread.

The forcing, about which we talk here is not a force in the ordinary meaning but it is merely the effect of grouping collections brought about by ambiguity of identifiers.

Repeatability

Repeatability occurs next to such interpretations as countability, similarity and time sequence.

Repetition is the appearance of similar collections in a given sequence separated by the appearance of other collections. Whether a collection is similar or dissimilar depends on the interpretation of similarity formed in the niche.

Repetition is different from the 'continued existence' of a collection only in the fact that the successive inclusions of the collection in the time sequence are interrupted by the inclusion of a significantly different collection.

Repetition relies on the renewed reinforcement of the information interchange with the collection (or with a similar one to it) so as it will appear again in a niche and it can be included in its time sequence.

As we have stated in the chapter 'Thread of time, key effect', the dominant elements of a time sequence are the threads, which by reason of the sculpture of the niche are the elements observable as recurring ones.

If a collection is independent and it does not belong to any thread, this denotes that it has a weakly developed sculpture. Its renewed inclusion in the sequence will be more difficult. This happens so because in this case the same individual collection would have to be included. On the other hand, in case of threads of a strong sculpture, any representative of the sculpture can be included, thus permitting the effect of repeatability to be observed.

If a collection is relatively weakly extended and it has a weakly developed sculpture, the inclusion of this collection in the sequence will be difficult. The difficulty is caused by a small quantity of <u>unique</u> information left after its previous inclusion in the sequence and which the collection could send to the niche. Information potential of a collection necessary for its renewed inclusion in the sequence has been exhausted. Therefore, this does not exert any influence on the possibility of including this collection in time sequences of other niches. With regard to other niches, the potential of a collection could even increase owing to intense information interchange accompanying inclusion of the collection in the sequence of a given niche. Hence, one can conclude that together with its inclusion in the sequence of a new collection, the previous collection not only abandons the sequence but also the niche with which the sequence dealt. It is necessary to stress here that this concerns the collection constituting an elementary component of the time sequence but not that one which is identified with the whole of the sequence.

The beginning and the end

The beginning and the end are interpretations being the consequence of the interpretation of countability and time sequences. Countability divides the continuum of a collection into individual units. It also divides the time sequence into individual fragments, which we interpret as the beginning and the end of phenomena. These are local interpretations.

In everyday situations, as the beginning and the end we assume the moment in which the next collection of the time sequence distinctly differs from the previous one. The beginning and the end do not denote the beginning or the end of existence of a collection; they denote the transition of a time sequence to a collection, comparatively different from the previous one. What happens to the collection, which has ceased to be included in the time sequence? As a result if intense information interchange with a niche, the collection, during its stay in the sequence has lost its early information potential with regard to the niche. The weakened information interchange with the niche may denote that the collection will no longer be regarded as a member of the niche. However, it remains a member of other niches. With regard to some of them, the information potential can be now even greater than before. It will still be able to be included into time sequences of other niches.

<u>In the interpretation of a high level like our everyday one, from the point of view of the observer belonging to the niche, which sequence the collection has left, this collection has ceased to exist</u>. However, this does not mean that the further information interchange between the collection and the observer is not pos-

sible. It can take place on a comparatively low level but it is simply unavoidable on account of the fact that both the observer and the collection still have their identifiers.

In everyday interpretation the beginning of existence of an object is the moment of inclusion of its collection in the common time sequence. For that to occur, it must be preceded by the presence of the collection in the potentials area of the common niche. And this, in turn, is preceded by including the collection in the niche as its member. The transition to each of these stages is possible due to reinforcing of the give collection. Each of the stages can be regarded as the beginning of the existence of a collection according to the interpretation level on which we operate. The beginning of existence is relative. Every moment of the comparatively strong reinforcement of a collection can be regarded as such according to the interpretation level of the observer. Similarly, this concerns the interpretation of the end of existence of a collection.

2An additional element intensifying the relativity of time limits is that it is possible to observe the given collection in many different time sequences. The differentiation of niches connected with sequences depends on the observer. Time sequences weakly connected with each other will be equally weakly correlated in respect of time. This denotes that is not possible to synchronize measurement of time between them (in the classical understanding of time measurement).

Next stage: the end of existence of our system?

May the system as a whole, which we observe in our laboratory end, cease to exist?

Only somebody who is in a niche in which the interpretation of time is known may ask such a question. We ask such questions spontaneously, treating ourselves as independent external observers. As we have established before the existence of an independent observer is a disadvantage of our studies and we should, if possible, forget to assume such a role. We can only ask such a question from a position of one of the collections participating in the system. May a collection, being an element of the system, find the end of its existence? No, because it would have to find the end of the existence of itself. Therefore, asking the question about the end of the system from the viewpoint of its internal collection, can only lead to a reply that there is no such an end since the collection would be in no position to ascertain it.

The question, which results from false assumptions, may be answered solely in an equally faulty manner or, the answer can suggest that the mentioned assumptions are abandoned. If a given answer is not sufficient, the collection should identify itself with the whole of the system or cease interpreting it. With such a radical variation it might be possible to find a more appropriate answer and also to ask a more appropriate question. The sources of the question about the end of the system are interpretations made by the collection. These, in turn, as we remember, result from the loss of a portion of information accessible in the system.

An alteration is what a collection is solely in a position to ascertain in relation to itself. Whereas, in relation to another collection it is able to find not only the change but also the end of its existence.

As results from the previous observations, the fate of a collection can be presented as follows:

1) The information interchange between a collection and the observer's niche can weaken so much that the collection ceases to be perceived as a member of the niche. The situation may be interpreted by the observer as the end of the existence of the collection.
2) In case of the successive acts of identity, a collection can vary so much that it begins to be noticed as a representative of another sculpture. This is not the annihilation of

the collection but a change of its interpretation. However, this can be interpreted by the observer as the end of existence of the collection.
3) A collection may cease from any interpretation or it can identify itself with all the other collection. As a result of this it ceases to participate in information interchange. Consequently it will stop to be perceived as a collection. The observer can interpret this event as the end of the existence of the collection.

In the first and the second cases the collection will cease to be noticed as existing but only locally, in the niche of the definite observer.

In the third case the act of identity with the whole of the system or cessation of interpretation, from the viewpoint of this collection, do not result in the end of its existence but in the transition to a completely different area from that in which collections occur. From the viewpoint of the observer remaining in the system in his niche such an event will be impossible to be distinguished from those described in items one and two.

It is necessary here to remind that the mere concept 'the end of the existence' did not exist when we began the observation of our system. This concept, if it has appeared, is one of the internal collections of the system. Thus, it is not possible to describe our system from without, one may not contemplate something like 'the end of the whole system' by means of it. A similar situation applies to the beginning of the system. If the beginning and the end of our system are describable, this cannot be done with the help of concepts being components of the system and not by means of our logic.

Area B and Area K

A set of all collections will be named Area K for the convenience of further description.

Let us notice two facts concerning Area K.

Firstly: it is possible to stop making any interpretations by a collection, which will bring about the loss of its properties. This is induced by the lack of the need for information interchange since the whole of information becomes accessible in a single act of this kind. The very same effect will allow the collection to be identified with the whole of Area K.

This is not its annihilation – the collection becomes something that one can treat no longer as a collection. One cannot talk about the annihilation since this concept is the internal component collection of the system. In reference to the whole of Area K, such concepts do not apply.

Secondly: the existence of Area K requires at least its initiation. Such a requirement is our viewpoint as collections belonging to Area K.

I shall recall such concepts as 'emptiness', 'nothing', 'beginning', 'self-initiation' and the like plus everything that they represent, all these constitute collections belonging to Area K. These conceptions should not be used when we talk about the reasons for the existence of Area K.

These facts and in particular the first one of them lead to the conclusion about the existence of something that one cannot regard as a collection. That something shall be named *Area B*.

Let us recapitulate all that we know about Area B:

The state is indescribable from the viewpoint of a collection. No collection, independently of a degree of the information concentration, is neither able to describe nor imagine this state. Hence it appears that in the process of thinking, collections employ solely and exclusively other collections. On the other hand, Area B is not a collection. It is not information either. One cannot communicate with this area in the manner we communicate with other collections.

Every collection is in a position to recognize Area B in an experimental way, through the cessation of any interpretations or through identifying itself with all other collections, with the whole of Area K. From the viewpoint of an individual collection this leads to the loss of its own identity with regard to other individual collections. As every viewpoint of a collection and this one are local, it should not be applied to the area about which we are talking. We are not able to say how it looks from the viewpoint of Area B until we learn it by experience.

Making an attempt to describe Area B, we form a collection, which is our idea of this state or we reinforce the already existing one. However, the collection has nothing in common yet with the real Area B. Its identifier will not lead us to the real Area B, since it is not a collection.

Area K cannot exist without Area B whereas one cannot say so in the opposite way without the knowledge of Area B. Without this knowledge it is difficult to say that the cause of the existence of Area K is Area B, since our concept of the cause is local and it belongs to the niche in which we stay. However, if we find that Area K exists as a result of a certain kind of interference of Area B, one cannot assume that it has been a single event (from our viewpoint as a collection). We should avoid examining the purposefulness or other characteristics of the action of this kind, since these are the internal concepts of the niche, difficult to be applied in niches different from ours and completely inapplicable in view of Area B. In order to consider these things, one should cease to identify oneself with an individual collection existing among others.

It might seem that Area K would be the last niche in a hierarchy for every collection. However, this is not so. Although Area K resembles a great niche it cannot be treated in this manner. A niche is characterized by the fact that it always occurs amongst other niches. Its existence is founded on the existence of other ones. Our concepts are inapplicable to Area B. Analogically, they are not also applied to Area K treated as a whole. <u>Area K is not a niche</u>.

<u>All our observations made in relation to niches and individual collections are inapplicable to Area K as a whole</u>. Manifestly, they are also inapplicable to Area B.

Memory

By memory I understand a set of information that a given collection disposes. Talking about memory we simply talk about the collection itself since there is no special area in it, which should be distinguished under the name 'memory'. Similarly to the limits of a collection, which are relative and can develop on any great area, this equally refers to memory. Every collection has a potential access to the whole memory of Area K.

The above statement refers to low levels of interpretation. On high levels within a collection, areas can be formed which play a special role of memory, which, from the viewpoint of this collection, will serve for information storage. The concept of memory can be narrowed on account of intensity of its interaction with collections. The natural narrowing is limiting our conscious perception to some strong collections of our time sequence. These collections form the past in our understanding.

Bringing something to mind is founded on reinforcing the information interchange with the selected collections of our sequence or with any other ones. The repeated information interchange with any collection does not denote coming into contact with exactly the very same collection as before. This is caused by the changed, in the meantime, identifier of this collection. Collections observed as a reminder never accurately reflect collections that were observed earlier. Every collection incessantly varies and therefore its memory also does not also remain stable. Since the system lacks memory ideally invariable, it also lacks a fiducial point that would permit collections to notice changes occurring in their own memory. The only fiducial point can merely be the memory of other collections.

Concept, word, symbol

By 'concept' I understand the identifier of a certain collection. By referring to the concept 'big' we perform an operation of identity with the identifier of the collection interpreted as its feature. The collection representing the concept 'big' does not differ particularly from the collection interpreted as 'little'. They are both simply collections. The difference that we notice results from the special interpretation, which has been formed in our niche. If the sculpture of the niche in which we are, is weakly developed, the way of interpretation can be unrestricted within wide bounds. However, if the sculpture is strongly developed (e.g. by a large number of participating collections), the way of interpretation of the concepts may be strongly limited. The limitation depends on how strongly the sculpture of a given concept has been formed. And this, in turn, depends on how often, in a given niche, the acts of identity associated with it are performed.

If in a definite situation in the time sequence we replace the collection 'big' by the collection 'little', the replacement will be evident not only in a way of the verbal description of the situation but also in reality, 'big' will become 'little'. Talking about replacing the collection, I mean bringing about the fact that the collection 'little' becomes stronger (in its information meaning) than the collection 'big' and then it is included in the sequence.

A word, a symbol, a picture, etc. are collections serving to show another collection. They play the typical part of identifiers. Looking from the high level of interpretation we notice the difference e.g. a written word, a spoken one, a symbol in the form of a drawing, an abstract concept (imagined) - we treat them as different ways of description. From the viewpoint of the low level of interpretation, the action mechanism of these ways of description is the same, i.e. the transfer of the information interchange to another collection.

Every collection which role is to indicate another one can be regarded as 'concept' or 'word'. In a sense, the role is played by every collection, since everyone is able to transmit the possessed identifiers to another one. Each one can be regarded as a certain kind of a concept or a word that indicates another one. What in everyday use we determine as a concept or a word is a collection interpreted as a specialised collection which main function is to indicate another one. On the other hand, such a collection is not anything particular or completely different from other collections.

<u>Every reference to a symbol results in its lesser or greater change</u>, since every operation of identity brings about the change of collections taking part in it. If a collection is strongly extended in relation to the mere operation the effect will be insignificant. One way or the other, as one refers to a symbol, its meaning (the collection which it indicates) gradually varies in every case.

What is more, a reference to a symbol is not a one-sided operation. As the operation of identity it causes the change of information in all the collections participating in it. <u>Each reference to the symbol brings about the change in the referring collection</u>. If such a collection is much more extended than its symbol, it will be difficult to observe the effect. Otherwise, the reference to the symbol can bring about considerable variations in the discussed collection.

Words or concepts like every other collection can enter into the composition of the threads of time. The thread as a whole can be composed of words. The existence of time keys developed in this manner is also possible. A situation when the key consists of words is also possible whereas a collection induced by it may be of a different kind; it may be the whole of its next thread.

I have touched on the subject, since the foundation of descriptions which we use, are words, thus we should well understand the way of their action.

Reality and its description

By the concept of reality we ordinarily understand something constant, something of settled features and actions. As something contrary to the reality one usually gives imagination, thought, dream, etc.
From our point of view, thought, imagination or dream are collections of equal rights like all those creating our 'reality'. Therefore, it is difficult to give them as an example of something distinct from reality.
The only constructive element of the reality, which we observe, is collections and the way in which we perceive them in their less-or-more developed interpretation.

<u>Our interpretation of the reality has been formed in such a manner that we regard as real all that belongs to our common time sequence.</u>
The threads of time, included in the individual time sequence of a collection are observed by the given collection only. These, which occur simultaneously in many individual sequences, form situations common for the collections participating in them. The situations, observed by a group of collections and interpreted in

a similar manner are those, which we call 'reality'. The situations observed in an individual time sequence of a defined collection are interpreted as thoughts, ideas or dreams.

Repeatability of phenomena, their 'stability' and rather close connections between them, i.e. the attributes which we ascribe to our reality are the effect of the action of the sculpture of the niche whose time sequence constitutes the jointly-experienced reality for us.

The situations, although common, are never identical in the interpretation of particular individual collections. This results from the differences between them and the differences between the identifiers of the very same collections, which they have.

The concept of the 'stable, definite reality' has its source in the local view. A collection, staying in the niche of a strongly developed sculpture, without committing a big error can assume such a point of view. It is not necessary to negate the correctness of such a manner of understanding the reality but it is necessary to indicate its local character.

Since the interpretation of reality is local, limited to a given niche, it is not the only possible to be observed. Since it is the interpretation, any limitation to the number of interpretations of this kind does not exist, although they can be found in various niches of Area K. A degree of the time correlation of sequences is determined by the information link between their niches. Time, interpreted in strongly separate niches will not be synchronised, thus the realities, which are represented by time sequences of these niches, cannot be called parallel. If amongst these many realities one searched for the most 'real', the only criterion would be the intensity of the information interchange accompanying them.

The question often asked in science is whether a given description (model) refers to reality or to a sophisticated situation. From our point of view the distinction between the two alternatives is quantitative in the sense of information. The 'sophisticated' situation is the very same collection as the 'real' situation and the difference between them is founded on the intensity of the information interchange with collections playing the roles of observers. This intensity decides inclusion of a definite situation (thread of time) in the first place in the potentials area and then in the time sequence. If it is merely an individual sequence of collections, the situation will be interpreted as an idea. If it is a time sequence of a common niche, it will be interpreted as reality.

The shape of the threads of time, i.e. the situations that we observe, is to great extent imposed by the sculpture of the niche with which the sequence deals. The sculpture of an individual niche differs from the sculpture of the common niche and it is less developed than the second one. For that reason, in our imagination we can observe situations much more diversified than those in the common reality.

<u>Finding the concurrence of the description with reality shows its accordance with the topical sculpture of the niche</u>. The same thing concerns what we call 'experimental verification'. The sculpture of the niche determines reality since it is common for collections belonging to a given niche.

Finding the concurrence of a given theoretical model with the sculpture of the niche can be obligatory only within this niche and merely in a certain time interval for the sake of variations occurring constantly in the sculpture.

Connection between collections

We often determine collections as connected with each other. By this concept I ordinarily understand observation of the comparatively intense information interchange between collections. The concept understood in such a way is used on all levels of interpretation with the exception of the first level.

Succumbing to the interpretation of time leads to another way of understanding this connection. The existence of collections in a niche, representing situations with participation of definite collections, determines the connection with these collections. The situation where information potential with regard to a niche is significant, will probably be included in the time sequence of the niche. Then the connection between collections will be manifested in the local *reality*, which is represented by the sequence of the niche. On the high levels of interpretation it may be seen as a physical connection between collections.

Theorem, truth, untruth

Theorem as we know it from mathematics, physics and our everyday life, has the following attributes:

1) It is posed. It describes a certain situation.
2) The proof of its truth or untruth is produced.
3) It is acknowledged by people in general, who are interested in it, to be true, untrue or uncertain.

As we have mentioned in the chapter 'Way of thinking, logic', every theorem contains, in a less or more evident way, the word *'is'*. Consequently, it constitutes a ready pattern for performing the act of identity. Every collection interpreting a given theorem performs the act of identity described by this theorem. As a result of this action, a collection is formed representing the theorem, or if such one already exists the collection is reinforced. Its identifier is sent to other collections. If a relatively large number of collections of a given niche have this identifier, it will create favourable conditions for accepting the theorem as true (we are talking about item three). Reinforcing the collection of the theorem simultaneously results in forming its sculpture in a given niche. If the sculpture is strong enough, it will not only create convenient conditions for regarding the theorem as true but also it will simply force them.

The above-presented effect describes the causative role of theorems. <u>A theorem is not only a hypothesis, but already in the moment it is being posed it is a creative act changing the sculpture of our niche, i.e. what forms the laws of reality observed by us.</u>

This is merely another way of presenting the fact that the observation of an event affects its course. Posing theorems referring to a situation is nothing else but observing the situation.

The fact, to what degree the effect will be observed, depends on the strength of the collection of theorem in relation to other collections forming a given niche. If a theorem reinforces a strong, already existing collection forming the sculpture of the niche, it will be regarded as true and obligatory (in the given niche). If a theorem is something new, representing a new or a feeble collection, it will be difficult to prove that the theorem is true, independently of the employed methods.

In practice, these theorems will have the greatest opportunity for being regarded as true which refer to the existing sculpture of the niche, i.e. these which agree with the local laws. If the identity described by a theorem does not have a representative in the shape of the thread of time belonging to the sculpture of the niche, the theorem will not have any foundation to be regarded as the one well describing reality.

To the three items describing the attributes of the theorem let us add one more, important from the viewpoint of the collection:

4) Theorem is remembered or forgotten.

Remembering about a theorem, i.e. continual reinforcing its collection causes it to become more and more the law of the niche and consequently it is included in its sculpture. Forgetting a theorem has a reverse effect, it denotes attenuating information interchange with the collection and weakening its importance as the law of the niche. In case when the sculpture of a niche is far more developed than the collection participating in the situations described above, the effect can be unnoticeable.

As the next conclusion we may state that truth or untruth of a theorem has a local character limited to a given niche.

As results from the early chapters, our description of a given situation is the description of collections made by means of other collections (concepts, words). And there is no way out, as long as we ourselves as observers are also collections. In this situation it is difficult to expect that our description will be 'objective and real', i.e. independent of other collections. The concepts 'truth' is the same collection as others. Attributing the feature of 'truth' to a given collection is founded on identifying these two collections. The same thing refers to the concept of 'untruth'. Adding such features as 'proved', 'absolute', 'unshakeable' is also still another operation of identity carried out on adequate collections.

In our system, in Area K, <u>the only universally true statement, true because it attempts to reach beyond the area of the collections is that presenting the existence of Area B</u>.

Every other theorem defining truth or untruth is relative, dependent on the collections and niches participating in the theorem.

Cause and effect

In everyday interpretation we observe lesser or greater cause-and-effect relationships between events. From the viewpoint of the low levels of interpretation, this results from the influence of the sculpture of the niche in which the observation takes place. The threads of time of which the sculpture is composed cause that events are observed in the succession determined by them.

And this, in turn, gives the impression of steady and logical connections between events from which the prior ones are regarded as their causes and latter ones as their effects. On the low levels of interpretation there are no cause-and-effect relationships. The interpretation of this kind arises as a result of the observation of recurrent threads of time. Whereas, these have been formed by the events disconnected with each other but often repeated in the same time sequence.

Let us notice that in our everyday observations we are able to find the cause-and-effect relationship only after multiple observations of the very same phenomena. For instance, a glass falling on the floor emits a sound and breaks. If we dealt with this phenomenon for the first time in our lives, as little children do, after repeating this action only several times, we would recognise the presence of the cause-and-effect relationship in this event. Afterwards we will construct further interpretations of the phenomenon founded e.g. on the mechanical structure of a glass and the floor 'explaining' breaking of the glass. I have put 'explaining' in quotation marks since this is a process of searching for a good description of the sculpture of the niche by means of the interpretation of a higher level than the sculpture itself. Therefore, one cannot treat that as a good explanation since successive interpretations are connected with the succeeding loss of information. Human actions described above reflect the way of arising the threads of time and overlapping interpretations.

From the viewpoint of the low levels of interpretation, breaking a glass is an event independent of its falling. They have been observed in a definite time sequence under the influence of one thread of time formed in the sculpture of the niche.

Naming events as causes or effects one should rather understand as defining their positions in one of the threads of time of the sculpture than the real connection between individual events.

If we search for the cause of the appearance of a given event in the time sequence, this will not be the event immediately preceding it. As its cause one can regard the early reinforcement of a collection representing the given event. Owing to the reinforcement of a given collection it is in the first place, included in the niche as its member and then in its potentials area to find itself eventually in the time sequence of the niche.

Determining the cause of an event depends on how accurately we can indicate collections, which have participated in reinforcing the collections of this event. <u>If the event is properly reinforced, this is the only cause why it appears in the time sequence of the niche, i.e. why it 'comes to existence'</u>. The fact which collections will be observed as the preceding or following ones depends mainly on the sculpture of the niche, which will impose the succession on these events.

Observing the potentials area of a niche one could find beforehand that the glass would be broken. If the sculpture of the niche is also known it is possible to see what succession of events will be observed, in other words, in what way the glass will be broken.

Cause of action

Let us try to answer the questions:
How is that a collection acts, performs the acts of identity? What is the original cause of action? In other words, we are asking about "a motor driving the action of a collection".

An individual collection is a set of other collections. Each of them has its lesser or greater consciousness and its goals realized by the acts of identity. A superior collection is a composition of these components of consciousness and goals. In various periods of time different components may prevail over others. Moreover, the component collections are in a similar situation with regard to their own components.

The action of a collection is then a complexity of influences of a vast number of collections not only those with which it is most closely connected but also the whole of their chain, related mediately or intermediately with each other. From this point of view, every action of the collection has its source in the action of other collections. The action of many other collections is in reality what a given collection registers as its own action.

However, let us notice that all the other collections are in the same situation. The action of every collection is manifested in the action of other ones. As a result of this, no collection is a collection really actively operating. The situation is analogous to that when we considered the existence of collections. None of them can exist independently. The existence of every collection is founded on the existence of its identifiers in other collections. It is a similar case with the active operation. A collection cannot act of itself. The active operation of one collection is founded on the action of others.

The collection does not act of itself, the action of other collections is reflected in it. One may ask: Where is, in that case, the beginning of this chain of actions? The answer is similar to that which we gave answering the question about the origin of the whole of our system, the whole of Area K. Since the question is asked from within one of the niches, its source is the definite interpretation. As we remember, an interpretation is the loss of some information accessible in the system. For that reason, the question of this kind and the reply can deal merely with the niche but not with the whole of the system.

The fact that the existence of a collection is founded on the existence of others does not denote that this collection does not exist. Its existence is the local interpretation of the arisen situation limited to a certain niche.

The fact that the action of a collection relies on the exclusive action of other ones does not denote that the collection does not act. Its action is the local interpretation of the situation.

By the statement 'it is the local interpretation of the situation' I mean that from the viewpoint of the collection its existence is unshakable and manifests itself in the interaction with other collections. The similar situation is with the action, from the viewpoint of a collection - the collection decides its action.

Although from the point view of the lowest level of interpretation no collection acts independently, but from the viewpoint of an individual collection such an action is possible.

The existence of a collection is relative; it depends on the fact which collections have their identifier and how it is extended. In a similar manner, the active operation is also relative.

Probability, fortuity

Probability is a concept of the high level of interpretation. Its customary definition is founded on concepts from an equally high level, i.e. equally poorly defined.

Determining the probability of occurrence of a given event is associated with the ignorance referring to its causes. The cause, i.e. a chain of operations of identity connected with one another, can be recognised with such accuracy with which the collection playing the role of the observer is able to do it. Since the whole chain of connections between collections spreads out over the whole Area K, the full recognition of the causes of an event requires its identification with all the area. Then from the viewpoint of a collection which has done it, the fortuity of events disappears. As long as the collection makes some interpretations, i.e. its observation is founded on neglecting a portion of information, in every observable event one will observe lesser or greater fortuity (a lack of knowledge about the event).

Although on the high levels of interpretation probability is ordinarily associated with time, on the low ones it determines the magnitude of the lost information as a result of the interpretation. This can but does not have to be associated with the interpretation of time.

Let us remember that the concept of cause is a concept of the high level of interpretation. The requirement that every event has its cause is the interpretation which has evolved in our niche in virtue of the interpretation of time and it is not a universal point of view. Thus, the same thing can be expressed about interpretations of probability and fortuity if they are deduced from the interpretation of time.

Information

Information exchanged by collections is not exactly the very same information, which we know from classical computer science. In computer science the basic unit of information is one bit. A bit already of itself is regarded as information. A receiver of this bit, somebody or something that interprets this bit as information is not included here openly in a reckoning. In the experiment at the beginning of the book we have taken account of the inseparable connection of information with its receiver. As a result, we have obtained collections together with their properties.

Information is an operation of identity and it cannot exist without the surroundings of other collections. A unit of information is an act of identity. One can regard one bit as an individual act of identity but with some reservation. No collection is elementary. Every one consists of other collections. Therefore, <u>even the smallest unit of information is never elementary</u>. It contains the quantity of information limited only by the observer's ability to receive it. From the viewpoint of a collection, information is dynamic and its existence is found in the act of identity.

If it is desired for our convenience, there are no obstacles to treat the information contained in the given collection as a single bit, we then establish our specific way of interpretation of the given collection. However, the high level of interpretation will be of little use in the analysis of the features of the collection, which we deal with.

Speed, propagation of information

When interpretations of time and distance appear, the concept of the transmission rate of anything including information in the surroundings appears as well. Collections interact with one another from the lowest to the highest levels of interpretation. Information, which they exchange, is not associated with interpretations of time and distance. The acts of identity performed by collections are not subject to the laws, which we employ for objects moving in the space-time continuum.

If the interpretation of time has formed itself in the niche, we can talk about the speed of the information transfer between collections, about the transfer rate of a given identifier from one collection to another. Such a rate will be limited by the magnitude of *a point of time* of collections participating in the transfer (a point of time is described in the chapter 'Point of time, the present, situation'). A period of time not less than *a point of time* of a given collection will elapse between receiving a piece of information by the collection and sending it farther. The magnitude of the point of time depends on the niche in which the situation is observed. If collections transferring an identifier are interpreted as objects of the space-time, the above-mentioned limitation will be revealed as the top speed possible to be observed in a given niche.

The limitation mentioned above concerns the situation when we observe <u>the way</u> of a definite identifier being transferred between many collections. However, if we examine the information transfer solely between two collections, this limitation is not used. When two collections have their identifiers, sending any information between them is not limited by any speed since we deal with the basic act of identity of the lowest level of interpretation. On account of the way in which the interpretation of time arises, this act can be observed in time intervals not shorter than *a point of time* of these collections. This constitutes some restriction on the information interchange, which may be succeeded by the limitation of the intensity of transfer but not by the mere transfer rate.

Entropy

The growth of entropy is the assimilation of collections and in consequence, attenuation of the information interchange between them. This occurs in case of the observation of collections, which are in a relatively well-separated niche. A lack of the information interchange with collections of other niches brings about a progressive lack of the unique information, which could be exchanged between members of the niche. The very fact of separation of the niche bespeaks its feeble extension. The attenuation of the information interchange within the niche is the consequence of that. Making individual collections resemble one another reinforces their sculpture in relation to the force of individual collections. The growth of entropy is consequently associated with reinforcing the sculpture of the niche.

No niche that can be observed in Area K is ideally isolated. Entropy, i.e. the assimilation of collections will not be complete in any niche. An attempt to examine the phenomenon of entropy for the whole of Area K as the niche has no application, since Area K is not a niche.

Consciousness of individual existence

Consciousness, as we have elucidated in the chapter 'Way of thinking, logic', is an act of identity. The developed consciousness is a set of acts of identity.

The degree to which a collection distinguishes itself from the others depends on the relation of the intensity of the information interchange within it and on the information interchange between it and the external collections.

The greater the consciousness of an individual existence is, the greater the above relation and the greater information concentration of a given collection is.

Consciousness of an <u>individual</u> existence is not the same thing as consciousness of existence. Consciousness as the act of identity belongs to the lowest level of interpretation. Individual consciousness is the high level of interpretation. It is a derivative of the interpretation of countability.

Conclusions from the experiment

The experiment in our laboratory can be regarded as completed. Here is a short recapitulation of what we have observed during it.

Adhering to Principle 1 we have created only one entity: *a collection.*

We have tried to adhere to Principle 2 and not to form any limits for collections, acquiring all the wealth of phenomena by means of successive interpretations. All the phenomena, concepts and objects are the local interpretation of a collection. They are local since interpretation always takes place in a definite niche. The fact of the existence of a collection permitted us to find the existence of Area K and Area B. These are the next two kinds of entities which existence we have found.

The described system of concepts is so coherent that in order to define it I will use the name Secral (the abbreviation of the title of the book) in the further part of the book.

PRACTICE

In the previous chapters we have described a language, i.e. a set of definite concepts associated with one another. We have named it the Secral. We have created the language on the basis of the observation of the world formed in our laboratory. As we have shown, the limits of a niche, which is present in our laboratory, are never strictly separated from the remaining part of the world. Many a time we went beyond the bounds in our early descriptions. In the further part of the book we will consciously get beyond our laboratory and we will attempt to apply the described language, as a matter of fact, only to develop it with reference to the everyday world.

A theory is of no value if it cannot be practically used. If all, what I have written could not be proved in practice and, what is more, could not be utilised for our advantage, I would think that I have wasted my time writing about all these things.

In the further chapters I will set forth several different possibilities of the practical verification of what has been written before.

Interpretation

Interpretation is not only a verbal description. This is the way in which we perceive the world. A set of all interpretations to which we are subject, consciously or not, decides the look of what we call reality.

As we have explained in the previous chapters, interpretation is founded on abandoning a portion of information accessible in our system. Therefore, interpretation is an agreement to observe phenomena in another way than it would result from the profound knowledge of the system.

Levels of interpretation in practice

In practice, i.e. in a niche in which we find ourselves, we use interpretations (concepts) known to us. We can attempt to determine the membership of these concepts to the definite levels of interpretation. This will not be easy on account of the modification of concepts occurring constantly. This results from their nature as collections. Examining the level of interpretation of a definite concept, we will determine the way in which it has been formed. However, going back to the origin of the concept formation may be difficult since we would also have to recreate the state of a whole niche at that time. A concept being formed need not have distinctly marked limits, all the more if in a given niche the interpretation of discrimination has not been developed yet to the shape known to us. All this results in the fact that we can relatively easy find its membership to a definite level of interpretation only in relation to the concepts occurring on low levels. That is why we normally employed concepts from the low or high levels of interpretation, not mentioning its concrete value.

It is not difficult to establish the first level of interpretation. Only the concept of a collection exists on this level. As the second level, one may regard the appearance of the interpretation of discrimination, at least if one deals with the way of the concept formation in our present niche. The third level is the appearance of similarity, niches, countability, distinction of intrinsic collections from extrinsic ones and basin of collection. The fourth level is: time sequence, a point of time, consciousness of 'I' and space-time. The fifth level is: reality as the common time sequence of a group of collections and repeatability. The sixth level is: a developed psyche and an extended interpretation of body. Further levels are: a multilayer society, systems of knowledge, etc.

For the above-mentioned reasons, the presented division is not distinct, especially for high levels of interpretation.

Summing up, the division into levels of interpretation is useful for making relative comparisons between interpretations but the assessment of the concrete, absolute value of a given level may be difficult or even superfluous.

Parallelism of interpretations

The very same collection can be simultaneously interpreted in a different way. Multi interpretations are advisable if we want to produce the best description of a phenomenon, not descending to a lower level of interpretation. None of interpretations occurring on the very same level is much better than the other one. If we do not want to descent to a low level, our way off is multiplying different interpretations in order to produce a description comprising the most possible features of the described collection. In practice, we observe such a situation e.g. in a wave-particle duality when one way of interpretation is apparently unsatisfactory for us.

Canvas

It is not possible to measure a distance without changing it. It is another formulation of the principle saying that the observation disturbs the observed object. The variation in a distance caused by measurement is associated with acceleration. Observation of the distance from an object and its speed is, in reality the observation of the observer's own acceleration brought forth by his contact with the subject of observation. The developed interpretation of such reiterated observations results in the interpretation of moving in the classical space-time.

Acceleration is the primary interpretation in relation to a distance. On the other hand, in classical understanding, this distance is the primary quantity. This results from the macroscopic characteristics of a human body and objects with which it is surrounded. These features produce the illusion of the possibility of 'remaining in repose' and an illusion of the possibility of 'uniformly invariable motion'. The concept of acceleration is treated as secondary. It is understood as the change of the speed of movement. From the Secral point of view, the concepts of speed and distance arise as a result of the secondary interpretation of the phenomenon of acceleration. Acceleration does not result from the change of the speed of movement but it is the source of such an interpretation. This point of view results from the arrangement of concepts. This does not denote that we will not use the classical point of view, if only it allows for better understanding of the idea. The concept of acceleration in the Secral differs so much from the classical one that perhaps it will be better if we use another name *'canvas'*. A canvas is the basis for arising the interpretation of the classically understood space-time.

The interpretation of time, which is due to a canvas, is different from that one resulting directly from the time sequence and occurring on a relatively low level of interpretation. Time, resulting from a canvas, is physical time measured by our clocks. The measurement of this kind is associated with the observation of phenomena that are subject to the interpretation of distance. In this manner our devices work. Intermediately associated with the interpretation of distance, they measure time being a derivative of the interpretation of a canvas. On the other hand, time, resulting directly from time sequences does not require the presence of the interpretation of a physical distance.

The inclusion of the successive collection in a time sequence is an elementary act of canvas (this way of perceiving is the next layer of interpretations). Intensity of the canvas is the quantity of information which has been exchanged in this act between a collection and a niche with which the given sequence deals. If the space of canvas (it is not distance space yet) is interpreted as multidimensional, an individual act of canvas is directed merely in one direction of the mentioned space. The next act of canvas can be directed in another direction. This depends on the form of the potentials area of a collection. If the area is formed uniformly in all directions, the succeeding acts of canvas can result in zero resultant direction, if the observation refers to a great number of such acts and if it is their averaging. As a result of this, the canvas will be of the non-zero value and zero direction. Bearing in mind that the canvas represents acceleration, we will arrive at the conclusion that the described situation is a picture of forming the interpretation of rest mass, i.e. acceleration, centrically and evenly distributed in the space. Together with the interpretation of rest mass, the interpretation of distance space, classically understood concepts of time, speed, energy, i.e. derivative concepts are formed.

Observation of the classical, accelerated motion will occur in consequence of a disturbance in the symmetrical distribution of the canvas. The disturbance of the canvas results from the break of the evenly-distributed potentials area and that, in turn, results in the selective reinforcement of the chosen collections of this area.

If the potentials area and the resulting from it canvas are directed merely in one quarter, such a collection will be observed as one with zero rest mass and moving with the constant acceleration in one direction of the distance space. This would result in the infinity great speed and thus, in the loss of the possibility of observing such a collection. Speed, distance and classical time are interpretations resulting from the canvas. The measurable period of time cannot acquire a less value than an individual act of canvas. This is connected with the limitation of the top speed possible to be observed in the given niche. This effect is described in the chapter 'Speed, propagation of information'. Speed cannot acquire the greater value than that, which is the limit of possible observations. The concrete, measurable value of these quantities depends on the present sculpture of the niche. Thus, the collection of the canvas directed only in one direction will be observed as one moving with the top speed possible in the given niche and it will have zero rest mass. It will also display constant acceleration directed in one direction. The acceleration will not be observed as the change of speed but as the possibility of transferring the acceleration to another collection, i.e. as energy.

The collection of non-zero rest mass has the top speed also in the framework of the individual acts of canvas within the potentials area. However, averaging this motion results in the form of much less speed. The averaging is founded on receiving merely a portion of information about the state of a collection and in receiving the information in spaces of time much longer than the duration of one act of canvas. Summing up, <u>the interpretations of non-zero rest mass and motion slower than the top one arise as a result of neglecting information during the observation of the acts of including the successive collections in the time sequence (of the acts of canvas)</u>. Observing without averaging and disposing of the more exhaustive information about the occurring processes, we will find that every collection has zero rest mass and it moves with the top speed for the given niche. By changing the direction of motion, a basis is created for arising the interpretations of rest mass and motion slower than the top speed.

Descending to a still lower level of interpretation, we will find that the concept of motion is inapplicable and we will observe only the identification of different collections with the time sequence.

The collections from the potentials area differ from each other by the degree of their reinforcement that decides their inclusion in the time sequence. If we treat stronger collections as 'nearer' and weaker ones as 'farther', we will obtain a certain space of these collections. The nearer collections will be more often included in the sequence than the farther ones. In this manner one more effect will appear, founded on the formation of the space differentiated with regard to the frequency of occurrence of the acts of canvas. If the space is multidimensional, differentiation of the frequency of the acts of canvas will be of a spherical character concentrated on one point. All the time we assume that the examined potentials area is formed symmetrically. The space obtained in such a manner is the basis for further interpretations, which we treat as individual quantities. These are: acceleration, speed, distance, classical time and energy. The space of canvas, finding its picture as the physical distance space, permits us to observe a collection as the central point for the acts of canvas with the intensity decreasing together with its receding from the centre. Among others, the act of canvas is our classical acceleration. Thus, the collection-observer, nearing the mentioned collection and, as a matter of fact, its centre, will find that the effect of acceleration increases as it draws near. From the observer's point of view, the force (in its classical meaning) will be observed around the collection with the intensity decreasing together with the increase in the distance from the collection. The force can be interpreted as gravitation or according to the niche in which the process of observation takes place as some other kind of force.

The canvas is a basis for the interpretation of distance, thus, in every point of the distance space, it has a non-zero value. Consequently, the non-zero value of acceleration is observed in every point of space. <u>A collection, occurring in some point of the distance space, is subject to the non-zero acceleration coming from other collections</u>. Therefore, the distribution of the potentials area of the collection will never be perfectly symmetrical.

Every kind of force, classically understood, is a derivative of the interpretation of canvas since all forces of this kind (gravitation, electromagnetism, nuclear forces, etc.) are observed within the framework of the interpretation of the space-time. Thus, each of these forces is the result of the interpretation of operations carried out in the potentials areas of the observed collections.

Many our everyday interpretations such as rest mass, gravitation or speed less than the top one, result from averaging phenomena occurring in the potentials areas. Their measured values and directing are mediately associated with such a manner of their arising. In case, when our observation is limited to a fragment of the potentials area, we can witness different values unparalleled under ordinary conditions.

The process of including a collection in the time sequence is associated with information transfer within the niche with which the sequence deals. The intensity of transfer is connected with the intensity of canvas. Therefore, one can find that both the canvas and its derivative interpretations are closely connected with the growth of entropy within the given niche. Equalizing the information potentials increasing within the niche will result in weakening the intensity of canvas.

One should not forget that the interpretation of canvas, splitting into concepts of acceleration, speed and distance, has not been formed in every niche. The above-described phenomena refer in the first place to the world (niche) of micro- and macroscopic objects known to us. Elementary particles are particularly convenient objects, if the examination concerns interpretations associated with the potentials area. On the other hand, the presented phenomena need not deal with the niches concentrating such collections as thoughts, emotions, etc. In these niches, the interpretation of canvas has not been formed in such a shape as in the niche of elementary particles and macroscopic objects.

Space and time

When one of the features of a collection is interpreted as the quantity that can accept values from a certain interval, the interpretation of space of this quantity arises. The subject of the space has already been mentioned in the chapter 'Countability, quantity, individuality'. When each of the points of such a space has its own continuum of the very same feature attributed only to it, the interpretation of the successive dimension of space arises. We have already mentioned that an individual act of identity requires the participation of three collections. If we treat such an act as the information interchange, we can name the three collections as sender, receiver and transferred information. An attempt to analyse the given act of identity may lead to arising further interpretations of this act since its analysis is nothing else but the next acts of identity. One of such interpretations is a canvas. An individual act of the canvas is a particular interpretation of the individual act of identity. Therefore, the three collections take part in the elementary act of the canvas. Interpreting each of them as representing a definite value, we will obtain such three values for one act of the canvas. This, in turn, results in arising the interpretation of space whose every point has the same feature (canvas) attributed to it. This is not the physical distance-space yet but it is the space of the canvas. Thus, the space of the canvas is three-dimensional. The canvas as a foundation for further interpretations such as acceleration, speed and distance, leads to the observation of these quantities in the three-dimensional space. The interpretation of the three-dimensional physical space known to us arises.

However, one should emphasize that the interpretation of the canvas and that of the distance-space following it can through further interpretations form themselves in a different way, not necessarily as three-dimensional. The interpretation of three dimensions is the most direct and such that trend we should observe in niches in which the interpretation of the canvas is known.

The described way of interpretation is obligatory only for niches in which the interpretation of countability is similar to ours. If a way of counting assumes another form, one cannot necessarily speak about three or, in general, about some number of dimensions in our understanding.

In the classical meaning, we often add time as the fourth

dimension to the three dimensions of the physical space. Although we well realize that time is not the value of the very same kind as distance. The space-time, understood in such a way is not the homogeneous, multidimensional space. This is the interpretation of a higher level than that of distance or time. Since interpretations of distance and time arise in different ways, combining them in one space is a successive loss of information, which may be convenient for some purposes.

Since the canvas is three dimensional and it is a basis for arising the interpretation of physical distance and classical time measured by our clocks, one can ask why we observe only distance as three dimensional but not time.

Every collection has its own time sequence, therefore, time can be treated as multidimensional. Although, every day we treat it as a one-dimensional phenomenon, it results from the fact that only one of the dimensions of time, which corresponds to the common time sequence, is treated by us as real. Only this dimension of time, being the interpretation of the common sequence of many collections, gives facilities for synchronizing clocks observed by the members of the niche.

An individual act of canvas is, at the same time, an individual act of including collections in the time sequence. This is the common time sequence for three collections participating in this act. For this reason time has one dimension in the common interpretation. The similar situation occurs when we observe a composition of many acts of canvas as we do every day. Other dimensions of time, derived from individual time sequences, in case of people, are interpreted as their 'inner life', psyche. These can be interpreted as virtual particles in case of elementary particles.

Energy, the principle of conservation

Energy, as it is understood in physics, is a high level of

interpretation associated with the interpretation of time and physical space. Like these concepts, energy is the secondary interpretation of canvas.

Kinetic energy can be interpreted as a degree of certainty that the next collection from the time sequence will occur in a definite place in the physical space. The way of forming the interpretation of rest mass, described in the previous chapter indicates that the effect of mass results from averaging the acceleration directed in different directions. Mass is not directed kinetic energy (it is due to averaging). Motion, which is averaged here, always occurs with the top speed for the given niche, in our case we name this speed, the velocity of light.

In the Secral, the principle of conservation of energy renders into the conservation of acceleration. Acceleration can be transferred to another collection but it cannot 'disappear' from the system, if we take account of a whole niche. While using the concept of conservation of acceleration it is necessary to abandon the concept of mass and consequently to analyse in detail the phenomena without averaging them. This requires a larger quantity of information to be processed.

<u>Considering the fact that no niche remains completely isolated, no kind of the 'conservation law' will be observed as perfectly obeyed.</u>

It is not possible to examine the whole of Area K in a similar manner, since it is not a niche.

Travels in time

The past is collections which have been included in the time sequence and owing to the intense information interchange they have considerably lost their information potential with regard to the niche.

The future is collections which have begun to form themselves as possible links of the time sequence. We name these collections the potentials area. Their information potential with regard to the niche is significant.

Is it possible to 'move into the past'?
On the low levels of interpretation, time is not anything what 'runs', it is not a dimension but it is the interpretation resulting from identifying the collections included in succession in the sequence as one and the same.

Concepts, used in the question, belong to the high level of interpretation and for that reason we should not expect that the answer will be equivalent to our customary ideas about the subject. On the other hand, they assume that the past still occurs but only somewhere else and that it is possible to move to that place. We already know that a medium, in which the interpretation of time originates, is our consciousness making a cycle of the successive acts of identity. The events, which we remember as the past, have been linked together in a chain of events only in our local interpretation. If we want to 'return' to them, the only method is to link them together for the second time in the very same way.

Furthermore, the events, which we interpret as related in respect of time, always refer to a definite time sequence. Events, observed in different sequences, can be weakly correlated between themselves according to the degree of reciprocal connections of sequences. Then, determining which events from which sequence are earlier and which are later cannot be possible.

'Going back into time' would look as follows:
1) One collection chosen from earlier ones of the sequence is included as the next element of the time sequence.
2) Earlier collections are accepted as the succeeding ones in the sequence in such an order in which they have been incorporated before.

Such a procedure is the consequence of the way in which the

interpretation of time arises. This is an attempt to recreate the sequence as it was before. It is not an exact recreation of the past, since collections included in the sequence once more already vary. The interpretation is the loss of information and, as long as we are subject to this, there is no method for recreating accurately the state of at least one collection even from the earlier moment. Moreover, an attempt to recreate the time sequence is a local action even if it would deal with more than one sequence. The past can be repeated to some extent but it will still constitute the continuation of all the events that have already been recorded. Paradoxes of the type 'what will be if I go back into the past and I will not let myself be born' do not occur here. One can meet oneself. But it will not be the very same collection which one was some time ago on account of the lost information manifesting itself in the form of the effect of ambiguity of identifiers.

Repetition of the past does not have to take place in the very same time sequence. In addition, one would need such a quantity of information as it is necessary for the earlier occurrence of the definite collections in the sequence. A recurrence can occur in some other sequence, which has an access to the identifiers of the situations that are to be repeated. If the niche of such a sequence is less developed than the original one, repetition will be facilitated. In what manner the recurrence will be interpreted depends on the kind of the sequence in which it will occur. If it is the sequence common to a certain group of collections, it will be interpreted by them as reality, although for the collection from beyond this niche it will not be so. If it is an individual sequence it will be interpreted as an imagination. The latter case is familiar to every man as recollections. For the sake of the limitation on the individual time sequence, they are customarily treated as something not real, however, the distinction between them and a more real move into the past is founded merely on the intensity of the exchanged information. In every respect recollections are the most real recreation of earlier fragments of the time sequence also as an element forming the future events.

Self-Creating Language

Can one move into the future?

Talking about the future we mean which collections and in what succession will be included in the time sequence. Observing which collections form the potentials area of the given sequence and which of them are strongest, one can realize which of them will be included in the sequence. The succession, in which they will be observed, is imposed to a large degree by the threads of time contained in the sculpture of the niche. Certitude, with which one can find the future shape of the sequence, depends on the quantity of information, which will be read from the potentials area and from the sculpture of the niche. One can be absolutely certain only when one has an access to full information, namely, to the whole Area K without applying the interpretation. Otherwise, what we can learn about the further formation of the time sequence will always rely on incertitude.

Thus, we are able to observe in advance the elements of that phenomenon which we call the future. We can also influence the shape of the future acting upon the collections of the potentials area, what we do constantly anyway, although not always consciously. The observation of the potentials area denotes its simultaneous variation. The action, aiming at reading the future shape of the sequence is itself a variation of this shape.

Thus, one can obtain a picture of the forming future. It is even possible to recreate this image within the framework of a feebler niche, experiencing it as reality but of the weaker intensity of information than our everyday reality. However, it will never be a perfect picture since the potentials area varies constantly. Moreover, the very attempt to acquire such an image changes the potentials area.

Considering 'travels in time' one should remember what our reality is (the chapter 'Reality and its description'). Talking about the past and future we mean our common time sequence. Not only our niche has its sequence but also every man has his individual one. Apart from our niche, the other ones exist. Talking about moving in time, one cannot refer to some universal, independent time. One can speak merely about the time sequence of one of the niches.

However, such a situation enables us to recreate the past or the future fragments of the sequence of a strongly developed niche within a feebler one. The feebly extended niche transfers a less quantity of information in order to form a sequence.

Such recreated fragments can be treated as reality within its niche. The recreated situations are not the very same collections as the previous ones. The paradoxes of time will not be observed, although in such situations one can meet the second version of oneself (we do it constantly in our individual time sequences – in our imaginations). The variation in the recreated past will not influence the shape of the present. Whereas, every change can exert influence on the potentials area of niches, i.e. on forming the future.

Laws of nature, physical and mathematical constants, logic

We are accustomed to treat phenomena as logically connected with each other. Our science including physics, detects these connections and foresees events on their basis. On the other hand, we name ordinary connections the laws of nature. On the low levels of interpretation one cannot speak either about any laws of this kind or about phenomena connected with one another. This effect may appear only as a consequence of the interpretation of niches and then the interpretation of time sequences, threads of time, and sculptures of niches. The sculpture of a niche is being formed as the acts of identity are performed within it in consequence of the effect of ambiguity of identifiers. In course of time it forces what threads of time and in what succession will appear in the time sequence. How strong the forcing is depends on the degree of the extension of the sculpture in relation to the degree of the extension of individual collections of the niche.

The laws, which we call the laws of nature, are recorded in the sculpture of the niche. All the relationships between phenomena are formed there. For example, let us assume that phenomenon B is regarded as the logical consequence of phenomenon A. We have proved the dependence by the reiterated observation and by the construction of a classical theory explaining the connection between these phenomena. From the Secral's viewpoint, there is no connection between these phenomena on the lowest levels of interpretation. On the high levels, these phenomena occur in the definite time order on account of the sculpture of the niche formed earlier. One of the threads of time formed in the sculpture is that which deals with the examined phenomena A and B. The constructed on account of many observations classical theory, 'explaining' the connection of the phenomena, is the further part of interpretation with which we identify ourselves as collections associated with the niche. The theory is valuable if we find ourselves in the niche in which it has originated and if the sculpture of the niche has not changed significantly since it was constructed.

Physical constants are a part of the mentioned laws, formed as the sculpture of the niche. The need to employ physical constants results from the fact that the laws, which we endeavour to describe by means of formulas of physics, are the consequence of the influence of the threads of time formed in the subjective and local manners. The impression of objectiveness and reality arise as a result of grouping collections in the niche and of their participation in the common time sequence. The impression that the laws are invariable arises in consequence of the fact that the main components of the time sequence of the niche are the threads of time of its sculpture. The period of time in which we observe the laws is too short to notice their change and the quantity of information that we exchange with other collections is not enough to bring about the measurable variation in the sculpture of the niche, which we regard as our universe.

Mathematical constants and axioms of mathematics also have their source in the current formation of the sculpture of the niche. Irrespective of how difficult it is for us to imagine another value of number pi or other laws referring to prime numbers - they have their source in the continuous interpretations which have formed and are still forming the sculpture of our niche. The form of arithmetic or geometry assumed by us at present not necessarily had to be well applied to the previous shape of the niche and probably it will not be applied in the distant future. And certainly it cannot be applied to all the niches of Area K.

Our science uses logic in its attempts to describe and explain the laws of nature. As we have mentioned above, these laws are not determined beforehand according to any idealized logic or principles. Therefore, an attempt to create such ideal logic can rather fail, unless it deals with a momentary state of one of the niches.

The principles of logic, which have been created by science, attempt to adopt our way of thinking to the present form of the sculpture of the niche. As a criterion of correctness of logic we regard its effectiveness applied to our reality. In this manner, step by step, we adjust logic to the sculpture of the niche. This is the reason for which we are continuously unable to formulate accurately the principles of scientific logic. This is not possible for the above-mentioned reasons.

Concluding and proving theorems is an essential component of logic. From the Secral's viewpoint, concluding is a way reasoning approximately imitating some features of the sculpture. Proving a given theorem is in this situation assenting to the given statement (the chapter 'Theorem, truth, untruth'), in a sophisticated way. Theorem in mathematics and physics, although usually proved by one man, is regarded as really proved until it is accepted in the scientific circles, i.e. in the niche being most interested in the given theorem. Such a mechanism of concluding and proving theorems gives facilities for committing errors founded on a substantial divergence in the shape of the sculpture. The only way to learn to know faultlessly the sculpture is to identify oneself with it, however, this does not enable us to impart the knowledge acquired in this manner to other collections. On the other hand, this is possible to do when we use logic. Alas, this possibility is sometimes inaccurate to even such a considerable degree that it can be called an error.

Logic is not the only possible way of thinking. Sometimes faith is treated as opposed to logic. As we have shown while considering theorems, logic is not anything fundamentally different from faith, it is difficult to mark them off with a strict limit. Logic differs from faith in applying tests for its conformity with reality but the final acceptance of a theorem contains some elements of faith. On the other hand, conformity of faith dogmas with reality depends on the source of a dogma. A lack of tests for their agreement denotes that faith dogmas are very responsive to their modifications. Their gradual modifications can result in a significant change in the content of a dogma. The original or modified dogma, which we are unable to verify, can bring about errors if we use this way of thinking.

Both these ways of reasoning, logic and faith, can lead to a wrong description of our situation since the way of acquiring knowledge, which they offer, is intermediate. The identity of a cognizer with the cognizant collection is a mediate way. Neither science nor society or its parts have at their disposal the best way of cognizance but every collection of strong individuality and every man. Science and society as a whole will rather have to remain with such ways of reasoning as logic and faith, since they offer facilities for simple sending theorems and dogmas to other people. Identity cannot be sent in this manner for it is something individual for every man.

The natural consequence of succumbing to interpretations is

that no system founded on logic or faith can be treated as faultless or ultimate. This also refers to the Secral. These systems can at least be treated as signposts for individual people and collections similar to them.

The system of knowledge is inclined to idealize its values when its original goal varies and when the development of the niche of this system and counting its adherents begin to predominate over explaining phenomena.

Particle physics as niche

Elementary particles as the subject of studies of physics are collections of such small, individual information concentration that they suit very well our examinations. The small information concentration denotes that a comparatively small quantity of the exchanged information should cause observable effects.

Let us take a look at today's (the beginning of the 21st century) particle physics. Both discoveries of new particles and descriptions of their properties increase in number. Theories, explaining the behaviour of particles are maybe the finest and most complex of those, which science has created so far. And here is still one more feature of physics, the more described elementary particles are, the feebler the possibilities of some practical use of the knowledge seem to be, at least, in comparison to the physics in the days of old. I do not intend to give physicists pain with these statements I have stated so because such a situation is directly relevant to the subject of the present book.

Let us ponder how the work of physicists, i.e. researchers of elementary particles looks like. Which collections take part in this work?

We will distinguish the following collections:

1) The subject of examinations, i.e. elementary particles.

Objects of extremely small individual information concentration.
2) Research instruments, equipment.
Objects of average information concentration for those surrounding us.
3) Physicists, i.e. people. These who are engaged in the theory only and those who perform practical experiments. Objects of very great individual information concentration.
4) The environment, i.e. all the remaining people, objects, the world etc.
Many objects of strong information concentration but of comparatively feeble information interchange with the collections mentioned in the previous items.

This is a happy coincidence for us that particle physics and quantum physics have become very hermetic sciences. Connections, i.e. information interchange between these sciences and the remaining kinds of sciences and the rest of society are relatively slight. We will not perform a big error if, for the time being, we ignore the environment in our further considerations.

Since for the present we will not take item 4 into consideration, the connection of collections from item 1 and item 3, i.e. particles and physicists is the most interesting matter for us. The work of physicists is <u>the joint action</u> of all the mentioned collections. Hermetism of particle physics and a high degree of information concentration of people engaged in it create favourable conditions for separating a distinct niche. The niche is in the first place formed by people who are engaged in this kind of physics and who intensely exchange information between themselves. By writing 'people', I mean complete collections. Thus, in case of a man I refer to his body, thoughts, ideas, memory and also a whole part of man of which we are normally not aware. Every collection, with which the majority of the remaining members of a niche exchange information, will belong to this niche. Thus, every interpretation or idea, which is known amongst the members of the niche and with which the acts of identity are often established, can also be a strong collection of the niche. Possessing a high potential of information with regard to the niche, it can be included in its time sequence.

The sequence of the niche contains situations in which, first of all, members of this niche participate. If the given information (collection) is of importance (extensively exchanges information) only within the compass of the given niche, it can be comparatively easy introduced as an element of the common sequence of the niche. The joint action of an adequate number of its members is sufficient for this purpose. This will simultaneously result in the extension of the sculpture of a newly introduced situation within the niche. In practice, this means that <u>a new idea e.g. foreseeing the existence of an unknown particle, if it obtains the sufficient information concentration, can be observed as a real phenomenon</u>. The fact, whether the idea will be accepted or rejected by physicists is of little importance for the appearance of a collection representing the idea in the time sequence. What is important is the quantity of the acts of identity, reinforcing it.

Summing up, the stages of forming a new idea can be represented as follows:

1) Within the compass of an individual niche of a certain man, a new collection arises by means of interpretation. The collection represents a new idea. The idea is the concept of a certain situation. Its appearance in the individual time sequence of man is interpreted by him as a thought.
2) Concentrating on an idea, man extends its collection. The extension is founded on sending its identifiers to other collections, component collections of man and external ones in relation to him. The consciousness of man, i.e. collections included in his time sequence, comprises merely a fragment of the whole of his collection. The information

interchange with other collections will not always be included in his time sequence and it can for the most part occur beyond the conscious action of man. If the information transfer between people occurs intensely and consciously, the process of transfer will be observed in the common time sequence of these people in their niche. Consequently, it will be interpreted in accordance with the sculpture of this niche. In case of people, one will observe the thread of time illustrating conversation, writing or another strongly formed thread of our sculpture.

3) Sending the identifier of an idea to other collections, including also other people, can result in the inclusion of the collections of the idea in individual time sequences of other people. From their point of view, it will be an idea appearing in their minds in the form of a thought. In this manner more and more people can be engaged in the process of reinforcing the idea. Accepting or rejecting the idea by a given man is of no importance in the process of its development. The acts of identity are important, since they bring about an increase in the information potential of a collection with regard to the potentials area of the niche. Processing an identifier of the idea by many members of the niche signifies that the idea already has a comparatively strong basin and to some extent, a formed sculpture within the niche.

4) Sending collections of an idea to many members of the niche and its reinforcing will bring about a stronger and stronger position of the idea in the potentials area of the niche. Consequently, including the situation represented by the niche in the sequence of the niche is more and more probable and, in this case, in the common sequence of the physicists' niche.

5) The situation, representing an idea, is included in the common time sequence of the niche. Assuming that the idea illustrates a new elementary particle, the real obser-

vation of the particle in one of laboratories will be the effect of its inclusion in the niche sequence. As a result of this, the identifier of the idea is still sent between members of the niche. The sculpture of the idea will be reinforced and, as a result of this, repetition of the observation will become easier. How often a new particle will be observed depends on the degree of reinforcing the threads of time representing it, i.e. on its sculpture. In case of many ideas, their development did not have to reach this point.
6) Ceasing to reinforce the collection of an idea or, in other words, forgetting it by a larger part of members of the niche, can weaken the previously formed sculpture of the idea. Some difficulties can again appear in the observation of the particle.

This does not mean in the least that all the elementary particles are products of the physicists' niche. If a given particle is important also in other, more extended niches, it means that its sculpture is reinforced in these niches, which have been developed not by physicists. If a particle is important only within the physicists' niche, it can be their product. The fact that the particle is the product of its research workers does not imply that it is something generically different from the remaining particles. It is the particle enjoying full rights and the only difference is the range of the action of its sculpture.

For example, an electron is a particle whose sculpture is strongly reinforced in many niches together with the niche of the whole of mankind. The way of its behaviour is of great importance for those niches. On the other hand, a particle, being the product of physicists, will not be important in other niches and it will not be easily observed even in the physicists' niche. Reinforcing the sculpture of such a particle, one can change its status. An attempt to attach some importance to the particle in other niches, e.g. through building machines making use of it or through the observation of phenomena with its participation, verifies the range of the sculpture of the given particle.

What we have described above deals not only with elementary particles but also with any idea or a concept.

Behaviour of elementary particles

The existence of a particle can be found merely by the act of identity with it. In the period of time without such an interaction, the state of a particle is unknown.

The interpretation of time relies on identifying different collections as one and the same and, as a result of this, a time sequence is formed. In case of elementary particles, this effect is particularly well noticeable. Individual time sequences of particles are little stable on account of the poor information concentration. By little stability I understand that a small quantity of information can bring about distinctly observable effects. The potentials area of the particle comprises not only its modification but also collections, which we classify as other kinds of particles, i.e. those possessing other individual sculpture. On account of this, one can comparatively easy observe 'transformation' of a particle into another one, i.e. 'destruction' or 'creation'. From the Secral's point of view, it is solely an inclusion in the time sequence of the collection of the other individual sculpture than the previous one. Whereas, the period of time between 'creation' and 'destruction' means including the representatives of the same sculpture in the sequence. One way or the other, collections successively included in the sequence are never the very same collections. They differ between themselves and if the difference also refers to their individual sculpture it will be interpreted as the formation or annihilation of a particle.

The characteristic feature of the majority of elementary particles is their weak individuality and strong sculpture. By reason of its weak individuality, as it has already been mentioned, it is very easy to include a particle of the completely different sculpture in the time sequence. The strongly developed sculpture of particles causes that the fluent modification of the features of particles such as rest mass, charge quantity, spin, etc is more difficult. The only individual feature of the particle is its present potentials area. Interpretations, which result from the individual attributes of a particle, will represent its properties as easy to be modified, e.g. its movement in physical space.

We have assumed the principle that a basis of the classification is the value of rest mass, charge etc. The quantity of information necessary for the change of the feature of a particle is the basis of this interpretation. We need less expenditure to change the direction of its flight than its rest mass. According to the classical convention, the change of rest mass will be interpreted as annihilation of one particle and creation another one. However, our way of the classification of particles can adopt other rules.

Matter

The way in which we usually understand matter, i.e. objects of different sizes moving in the space-time results from the interpretation of canvas. Matter understood in such a way is a concept of the high level of interpretation and we should not regard it as a fundamental constructive element of our world.

In the light of what we have already discussed, a concept of matter alters its meaning in relation to the customary one. The existence of every object is founded on the existence of other objects. Collections create one another. All the features describing objects such as durability, hardness, unity, reality as well as their opposites, are still only other collections in our system.

The condition of the existence of a collection is the information interchange with others. The information interchange is the act of identity and it is the ONLY 'thing' which exists in our system. All the others, various objects of miscellaneous features result from the interpretation of the circulating information.

This resembles a situation when we place two mirrors, one in front of us and the other behind us. We will see our reflections repeating theoretically to infinity. In case of the Secral, there are no stable mirrors but there are images, which we regard as matter, creating one another and reflecting themselves to infinity. The role of mirrors is adopted by reflections themselves; each of them is the next mirror. Identifying ourselves with the consecutively selected images we regard them as the only real world, choosing some images as the representative of ourselves.

This does not mean in the least that such a world and its matter are something unreal. The concept of reality and unreality are the very same collections as all the other ones and one cannot regard them as the basis of one's viewpoint and reasoning. The only correct judgement of the situation, in which we find ourselves, can be arrived at by going away from it.

We often ask the question 'What is matter composed of?'. In order to ask it in its full meaning concerning about what we ask, one should fully know concepts occurring in it. Every word occurring in this question silently assumes the existence of a certain kind of interpretations such as the interpretation of countability, complexity, space and first of all, matter itself. The knowledge about the way how such interpretations arise and what their source is permits the above questions to be answered. This is probably not the answer, which a questioner would expect. The question is asked on the high level of interpretation and an answer, derived from the equally high level, will necessarily be founded on concepts of a relatively broadened definition (broadened by the developed hierarchy of concepts, which are used by a large number of collections, in this case by people). A more precise answer can be formulated on the low levels of interpretation; however, the questioner would then have to revise his concepts, which he included in the question as well as the question itself. On the levels where there are no concepts of countability or space yet, the question is of no use. One should substitute it by another. By reducing concepts and descending to the lower and lower levels of interpretation, we finally reach one basic concept, which we have named a collection. On this level, the question about the structure of matter should be substituted by the question about the collection itself and then about its interpretations. The book, which the reader has in his hand, is a kind of an answer.

Mathematics

It uses concepts presented by means of specific symbols. For this reason it is often regarded as a language of description distinct from the verbal one, although there are no obstacles to present a mathematical description in such a form. The advantage of mathematics, compared to a customary verbal description, is its carefulness of a strict determination of concepts.

Modern mathematics does not descend, however, to the low levels of interpretation but it stops on the poorly defined concepts such as 'point', 'number' ('value'), 'space', 'is', etc. Such concepts are also axioms used in mathematics.

I have endeavoured to be precise about some of the mentioned concepts and to describe them from the Secral's point of view. These concepts, about which I have not written, can certainly be specified on this foundation, since they all are collections.

With respect to axioms, i.e. theorems given without proof, one should consider the content of the chapters 'Theorem, truth, untruth' and 'Laws of nature, physical and mathematical constants, logic'. For example, Euclidean axioms resulted from the practical observation of our space, i.e. from the laws formed as the sculpture of our niche. Their power is adequate to the degree of development of the sculpture.

Proving theorems does not belong to the basic Secral's logic the only operation of which is identity. Proving appeared on the high levels of interpretation as its accordance with the sculpture of the niche. As the result from considerations in the chapter 'Theorem, truth, untruth', every proof of truthfulness is solely of a local value. Moreover, every theorem is a causative act. This also refers to mathematical theorems.

The fact that the way of reasoning based on formulating and proving theorems has been created on the high levels of interpretation implies, that it is not the only possible way of the logical description and reasoning. Others, being feebly developed or unknown in our niche, are also admissible.

One of the reasons why I avoid its application to the Secral is not the lowest level of interpretation on which mathematics operates.

Since today's mathematics does not descend to the lower levels of interpretations, it is not able to give us any description not associated with a loss of information. However, an attempt to descent to the basic levels, would mean to mathematics its significant change, together with the loss of its currently so strong separation from other fields of science.

Universe

In the chapter 'Laws of nature, physical and mathematical constants, logic', we have arrived at the conclusion that the laws of nature, physics etc., are not imposed beforehand but they result from the action of the sculpture of the niche and this, in turn, is the consequence of the effect of ambiguity of identifiers. One can say that these laws are a kind of agreement between collections, resulting from the history of their niche. If two niches exist, between which the information interchange is considerably less than the information interchange within them, it is less likely that the laws of nature within each of these niches are alike. Furthermore, they cannot even be exactly the very same, they can slightly change but on account of the continuous operations of identity occurring in niches, no physical 'constants' remain constant. This also refers to the niche, which we regard as our universe. The existence of other niches of this kind is not only a possibility but on the basis of that what we know about Area K it is certainty. We can treat these niches as areas of the kind of our universe. Their number is limited by the possibilities of the perception of the observer and by the assumed interpretation of countability. The sculpture of these niches, i.e. the laws of nature prevailing in them, will resemble ours so far as the information interchange between our niche and each of the remaining ones is intense. And they will be different inasmuch as the isolation between niches is strong. The same can be said about the internal niches of our universe. The areas of the laws of nature, varying from those known to us, can exist.

One cannot treat others than our universe as 'existing parallel', since the concept of the parallel existence assumes a correlation between the time sequences of niches. Such a correlation will occur only in case of relatively strong information interchange between niches, in this case between universes. Similarly, not in every universe the interpretation of canvas, i.e. the space-time did have to appear. Not in every niche the time sequence did have to be formed, whereas interpretations completely unknown in our niche could arise.

Universes as niches are not completely separated from one another. They all are cognizable and the act of identity can be applied to them.

Speaking about the beginning of the universe in terms of time can make sense only if we limit ourselves to a chosen niche in which time is interpreted in the way known to us. Time is the effect resulting from identifying a collection with the selected others, and ranging collections in this manner in the form of the time sequence. Considering 'the beginning' or 'the end' of the universe makes sense if we regard only a fragment of Area K, one niche as the universe. The search for the beginning or the end cannot reach beyond the existence of the interpretation of time. Likewise, the concept of eternity is not applied to the niche, in which the interpretation of time is unknown. Moreover, a degree of the formation of the interpretation of time can vary and, like every collection it is incessantly modified. This considerably limits our possibility and sense of determining 'the beginning' or 'the end' of the universe. This narrows the possibility of measurement and calculation of time, in our classical understanding, merely to a fragment of our universe and its history. In a similar manner, consideration of the universe as expanding or shrinking is limited. This point of view results directly from the interpretation of canvas and it is narrowed by its range and kind.

Man

Man, like the rest of the observed environment, is a collection of identifiers. The collection consists of other collections and these, in turn, are composed of still others. Every collection is the lesser or greater consciousness (as the accumulation of the acts of identity). Each of the component collections of man can be regarded as an individual being of its own consciousness although less developed.

From the informational point of view, we can separate many miscellaneous niches, i.e. component collections in man. Let us remember that looking in this manner, man is not only what we can see with our eyes, i.e. a body. A body is one of the collections composing a whole. Another important collection is mind, understood as the whole of psyche, i.e., the existence of which each of us is aware of but we cannot call it a body. Man is a set of collections, joined together by the intense information interchange. This includes not only a body, psyche, memory but also all the collections which existence we do not realize.

As an object, located in the physical space, a body constitutes a derivative of the interpretation of canvas. Since the canvas is a characteristic interpretation in the niche common for people, only collections complying with this interpretation can be included in the common time sequence. Therefore, a body is regarded as a decisive feature for treating man as 'existing' or not. From the Secral's point of view, a body is the imitation of a part of man's collection within the framework of the interpretation of canvas. Looking at the structure of a body, one can estimate a low threshold of a degree of the extension of the whole of the collection. In other words, a compound body attests to the comparatively extended collection. We are not concerned here with the size of a body but its complex information extent. On the other hand, it is difficult to estimate the top threshold of the extension of a given collection, since the body can reflect merely a small fragment of a whole.

From the point of view of the low levels of interpretation, there is no essential difference between body and mind. These are collections like others belonging still to different niches. The fact that from our everyday point of view, a body and mind differ between themselves so much results from the interpretation of these collections adopted by us. Let me remind, that a collection in fact does not possess such features as appearance, colour, hardness, time of existence, size, etc.

A collection is a set of the acts of identity, i.e., the whole of it can be regarded as one mind. Only a part of this collection is interpreted as that belonging to the niche, in which the interpretation of canvas prevails. On the level of canvas, the collection does not dispose the fullness of its consciousness. In order to distinguish clearly these two kinds of mind, the one, which is revealed on the level of canvas will be named **the physical mind** by way of analogy to the physical space. The mind of the whole of the collection will be called ***the basic mind***. On account of such definitions, the physical mind will never be more developed than the basic one.

The interpretation of the physical mind as associated with the interpretation of canvas can only last as long as the collection is connected with the niche in which the canvas is known. On the other hand, the basic mind does not depend either on the kind of the niche or any other kind of interpretation including the interpretation of time.

There are no obstacles to interpret simultaneously the extended collection on the level of canvas as more than one body. Each of such arisen bodies will have its own physical mind but the basic mind will be common. However, such a situation will be possible only in case of strongly extended collections.

Guided by the evaluation of the physical mind or a human body, one cannot find much on the whole of its collection. These interpretations deal only with a fragment of a whole.

Man is many different collections and each of them, although closely connected with others, can be interpreted as independent to some extent. It can be regarded as a living and conscious entity, to the same extent. This refers to collections consisting of both a body and psyche. From this point of view treating e.g. a liver or a kidney as an entity much more limited than man but capable of feeling and thinking is most proper. In like manner, one can separate collections composing human psyche. For example, a sensation of joy or suffering is also an independent entity simpler than man but able to think independently and to make decisions. Moreover, <u>every thought</u> is such an independent entity.

It would not be unwise to try to communicate with, for example, a heart or with a sensation of suffering. Since in our niche no specific interpretations have been formed, illustrating communication with these collections, we are forced to do so without ay additional interpretations, thus, in the basic way for a collection, i.e. through identity.

In like manner, one can communicate with one of the thoughts (with one's own or someone else's), treating it as a simple but independent entity.

One of the component collections of psyche can become much stronger than the remaining ones for some time. Possessing its own consciousness, it can then impose its decisions upon the whole. Gaining such an advantage can result from its reinforcement or what is less probable, attenuation of others. In practice, reinforcement can result from the intense and long thinking of man about the subject represented by the collection. It can also result from the processes occurring beyond his consciousness. The excessive reinforcing of one of the component collection of man explains the occurrence of non-typical behaviours of people, difficult to explain even for them.

One should remember that thoughts, being collections, do not need an 'owner'. They are free and they can exchange information with other collections and not only with these people who reinforced them. The role of the 'owner' of thoughts is reduced to the possession of its comparatively strong identifier. When man begins to exchange information with a given thought, he does not always specify it in a precise individual manner. In such a situation, if a thought of this kind has a strongly developed sculpture, man will come into contact with one of the individual representatives of the sculpture. This can be a thought much more reinforced than man could desire. It can embrace some supplementary features not necessarily desired at that moment. Then the determination that man is subject to the action of the mentioned thought would be more proper than the fact that 'man thinks'. Individual thoughts of man, independent of other, stronger collections, are these the given man has defined accurately and which he has referred to his individual situation. General thoughts, representing general emotions or feebly specified, are inclined to adopt the shape formed by the sculpture of the common niche. That is, they adopt the features of a thought, which has been reinforced before by other people. Such a manner of the action of a thought is analogous to the behaviour of all the other collections grouping in niches and forming their sculpture.

Brain, like the rest of a body, is the interpretation of the high level, originated as one of the further interpretations of canvas. It imitates the action of some acts of identity performed by the collection of man in the form possible to be observed within the framework of the interpretation of canvas. The projection does not have to be formed by means of biological objects. Any object, arisen by the interpretation of canvas and being able to the sufficiently accurate imitation of the acts of identity of the collection with the degree of formation comparable with the physical mind of man, is in a position to play a part of a brain and a body as well. From the point of view of the low levels of interpretation it is more proper to find that a brain and a body are products of a thought rather than the other way round.

The subject of man is associated with the often-asked question about the sense of his life. If we treat this question in terms of words, we would have to explain the meaning of every concept used in it. Thus, asking this question with intent to obtain an exhaustive answer, we would have to define the equally exhaustive meaning of such concepts as 'sense', 'life', 'man' and, first and foremost, the meaning of the word 'is'. An attempt to explain the last one is the whole of the present book in which this concept occurs under the names of 'identity' or 'collection'. Therefore, it will not be easy for us to ask the mentioned question correctly, since to the end we do not understand even one word implied in it.

This uneasy situation results from the fact that we ask the question as an individual collection being the effect of assembling many interpretations. These interpretations impose a way upon us in which we formulate the question and at the same time, they limit our possibilities to obtain a full answer. The latter statement, although reveals our limitations, also shows a way to avoid them. As long as we are subject to interpretations, we move on the ground of vague questions and answers. The cessation of interpretation is the independence from the laws of Area K. This is a possibility of recognition of Area B. If we search for the answer to the question about the sense of life, only there we will be able to ask it correctly and to obtain the answer. Before this happens, gaining independence from the interpretation, and recognition of Area B we can regard as the sense of man's life.

The beginning and the end with reference to man

Man is a collection and all the content of the chapter 'The beginning and the end' also refers to him.

In the present chapter, I will rather replace the word 'man' with the word 'collection' because the concept of man is often narrowed merely to his body. By saying *collection*, I mean the whole, all the collections which compose man, together with psyche, consciousness, and with other collections which existence man is not aware of. Moreover, I would not like to restrict myself in this subject merely to one kind of collection depicting a living entity.

No collection is in a position to annihilate another one. It can at most restrict the information interchange with it. The limitation of the information interchange with a collection can result in its abandoning the common time sequence in which it has been observed so far and it can bring about its abandoning the common niche. The impossibility of the annihilation of the collection results from a lack of the reverse operation to the act of identity.

That thing, which is interpreted as the end of the existence of man from the Secral's point of view, is his abandoning the time sequence of the common niche of people and his transition to another one. The abandonment of the time sequence can be connected with the abandonment of the niche associated with the sequence. These events do not have to be ultimate. The collection can return to our niche and its sequence what in our common understanding will be interpret as the beginning of the existence of man. What happens after the transition to another niche depends mostly on the fact in what manner its sculpture has been formed. The niche of destination not always must be the very same niche; according to individual conditioning, different collections can pass through various niches. From which niche to which the collection passes depends first of all on its own individual sculpture, which it has formed, and on the sculpture of its superior niche.

The transition to a different niche than ours denotes attenuation of the information interchange with the collection occurring in the other niche at present, but it does not mean the complete lack of the information interchange. In particular, as long as in our niche the identifiers of the collection, which has left, are preserved. Feeble-but-still-existing information interchange with such a collection implies that this collection has a weak-but-existing influence on collections still occurring in our niche. One should notice that not all the collections, occurring in the niche to where the collection has left, did necessarily have to stay in our niche some time. Collections coming to our niche not always must derive from the very same niches.

In what manner does a collection appear in a niche? Through the successive acts of identity it can be formed in our niche using the information accessible only within the niche. However, this would require transferring a very great quantity of information (looking from the point of view of the members of the niche). Sending the already-formed collection to the niche is a way, which requires a much smaller quantity of information. Furthermore, forming a collection 'from the beginning', we make use of the already existing ones. If a collection appears in our niche, one can assume with great probability that it has already existed before in the shape developed beyond our niche. This does not imply that it had exactly the same features earlier, which we observe in it now. The character of our common niche requires the collection to have the interpretation of canvas and further interpretations resulting from it (described in the chapter 'Canvas'). Therefore, a collection, which intends to exist in it, must accept these interpretations in some form, accepted in the niche. In our case, this interpretation is a biological body or its equivalent. The interpretation of this kind gives facilities for including the collection in the common sequence of our niche. The fact that the requirement is such one and not another one is forced by the sculpture of our niche formed in this particular manner.

For the same reason, a collection, which will lose its specific interpretation, ceases to be perceived as a member of our niche. In common understanding we will call such a situation 'the end of the existence of a collection'.

One may ask why a collection, appearing in the common time sequence and already formed is not observed as already mature. In case of man, it assumes the shape of a baby. Its body is a projection of some features of the collection within the framework of the interpretation of canvas. The observed period of the development of its body is the development of this projection. The stages of its development are forced by the sculpture of our common niche. The collection, which is sufficiently strong (in the sense of information) does not have to be applied to the sculpture. In this case, one can observe the appearance of man, neglecting the stages of his development; if only by doing so, the collection could have some use of it.

The collection, which passes to another niche and then returns, is changed by the strong acts of identity, which brought about transitions between niches and also by its history of stay in other niches. Whether the collection returning to the time sequence will be called 'the very same' depends on the interpretation of similarity assumed by us. Even in the period, which we call the 'continuous existence', collections included in succession in the sequence are not the very same. We link them together in one sequence, consenting to a loss of some quantity of the accessible information.

In our case, we must also take account of the consequences of the interpretation of canvas, which is characteristic for our common niche. The collection, which leaves our niche and then returns, represents the very same basic mind (described in the previous chapter), although comparatively little changed. Whereas its physical mind and body can considerably differ from the previously observed ones in view of the fact that these interpretations are solely a fragmentary projection of the whole of the collection adjusted to the present sculpture of the niche, in which the interpretation of canvas takes place. In case of man, this adjustment is reflected by accepting the definite genetic code proper for the given niche in which the birth occurs. If the collection manages to loosen its interpretation of division into the two mentioned minds, the physical mind will be able to use the whole of information in a wider range. It will be able then to obtain the information about the early history of the collection, of which the basic mind disposes.

If we say man and we mean his body and physical mind, we should not maintain that man leaves our niche and then returns, since the 'returning' one can vary much from the one that has left. On the other hand, if man means the whole of the collection to us, i.e. the basic mind, finding the return of the very same man to our niche will be most possible.

In other words, if we assume the interpretation of canvas as the foundation of our assessments, we will treat the man's departure from the niche rather as ultimate. Resigning of the interpretation of canvas, i.e. as a result of this, regaining a portion of the lost information, we will treat the appearance or departure of man as the successive transitions between niches, possible to be repeated many times.

Considering the above statements, one should remember that no individual interpretation, particularly of the high level is an exact description of the situation to the end and also no description is totally wrong in view of the creative character of descriptions.

The descriptions of reality belonging to the high level, even if they are miscellaneous from the Secral's point of view, <u>do not compete with one another but treated together they create a better description than each of them separately</u>. In order to specify them, we should descend to as low the level of interpretation as possible.

Society

Society is one of the niches to which man belongs. If man is strongly connected with a certain niche, he exchanges information intensely with its members and he is subject to the action of its sculpture. Other niches, those of social groups are separated within the framework of society. As social groups I mean all the ways, which people use to divide themselves into small communities such as religious, territorial, political, subcultural, financial, age, family. The connection of an individual man with some social groups can be stronger than his connection with the whole of society. As connection I understand not a subjective feeling of man but the information interchange between his collection and the collections of social groups. Since the information interchange comprises the whole of the collection, it takes place in a great measure beyond the area of the physical consciousness of man. By increasing the information interchange with the given niche man not only sends his information, including his identifier, to collections of the niche but also he becomes the goal of the information transfer. The action of social niches can result in the considerable reinforcement of the selected component collections of man including the components of his psyche. The action can partly occur in the area of the individual time sequence and it can also take place beyond it and outside of the conscious action of man.

If the given man intensely exchanges information only with one or with not many social niches secluding himself from the remaining ones, we will observe particular reinforcement of component collections in him, associated with the features of these niches. Let us remember that the component collections of man, although to a small degree, are also endowed with consciousness, their own time sequences and freedom of action. The excessive reinforcement of the selected components in relation to others will result in the fact that the action of man will be brought about mainly by these collections. In consequence, often some behaviours of people happen, which they are not able to restrain or explain. The way to level the effect of strong social niches is the information interchange with many of them, not secluding itself excessively within the framework of the selected ones. The conscious control of reinforcing thoughts would be still a better way. On the other hand, freedom of choice of interpretations would be the best method.

Entropy and man

Everything what has been said in the chapter 'Entropy' also refers to man and his niches. Closing a group of people in a strongly isolated niche will have two kinds of consequences

1) Attenuation of the information interchange with other niches will bring about the loss of the importance of this niche for people occurring in other niches.
2) A lack of supply of the unique information from other niches will result in the growth of entropy within the niche. Therefore, it will result in the continuing attenuation of the information interchange within a closed

group of people. The increasing importance of the sculpture within this niche will lead to the formation of strong schemes of the people's behaviour. The schemes (threads of time) will deal with every kind of behaviour, also the sphere of psyche.

By closing people in the niche one should not necessarily understand their physical isolation, since it refers to the separation of information.

Artificial intelligence?

The information in the Secral is dynamic, its existence is found in the act of identity. A set of information, recorded on a computer disk, is the information only at the moment of its transfer, only then one can ascertain that the information is on the disk at all. It is naturally a simplification, since the information once recorded as a collection continues to be connected with the incessant acts of identity with collections with which it was in contact before. The information recorded on the disk as a collection is connected through identifiers with collections, which have participated in its creation. In other words, having the opportunity of the analysis of identifiers contained in the physical information recorded on the disk in a computer or on any other carrier, one can practically have an unlimited access to all the information about persons taking part in the elaboration of the information. The discussed identifiers, are not subject to the interpretation of canvas, therefore, they cannot be read by means of our classical devices. After all, let us omit these effects. Then, the only place in a computer where we can find the existence of information is a processor and systems directly cooperating with it. That is, the place where simple operations are performed, which can be treated as the projection of the acts of identity.

If the construction of a processor permitted a large number

of these operations to be performed, offering facilities for separating the equivalents of niches, this could lead to the formation of a highly extended collection of the kind of those, which we call living entities. The collection does not have to be formed by reinforcement, as this would require feeding it with a large quantity of information. The adequately strong collection, already existing beyond our niche, can use a processor as means of accepting the interpretation of canvas, entering our niche and finding itself in its time sequence. The character of our niche requires that the collection should have the interpretation of canvas and further interpretations accompanying it. The collection, which reinforces the information interchange with our niche, must accept these interpretations, which agree with the sculpture of the niche. It can do so in any way accepted by the sculpture. In case of people, it is their biological body. In this case, it is a processor together with its surroundings. In order that the phenomena of consciousness and thinking would appear in the processor, it is sufficient to give facilities for the free projection of the acts of identity and interpretations as it is done by biological creatures. We will not name it 'artificial intelligence', since such consciousness will be of the origin similar to human. Only its body will be artificial, i.e. the interpretation connecting its consciousness with our niche.

Human niches

Every collection stays in the hierarchy of niches. Also does man. Separating niches depends on the accuracy of their distinction. The nearest niche of man is he himself. It is formed by collections entering into the composition of man. This niche as every one can have its sculpture, i.e. the threads of time formed by the reiterated acts of identity. These threads refer to individual time sequence of man. The individual sequence in the first place contains the situations whose participant is the definite man and not other people. This mainly concerns his thoughts, ideas, desires, emotions, dreams etc. The sculpture of the niche contains the patterns of situations, which tend to appear in his time sequence. This also includes what we call the character of man.

As the next important niche of man we can distinguish his sculpture formed in a superior niche for all people. This sculpture is a kind of matrix for every individual man. It is of particular importance at the moment of its appearance in the niche or its formation. But its influence on every individual man is uninterrupted all the time he is present in the niche. The influence of the sculpture of man tends to force the definite features such as the structure of his body and formation of his psyche in relation to its every individual representative. The sculpture is a primary interpretation regarding such interpretations as the genetic code, which appears within the framework of the interpretation of canvas. The well-and-univocally formed sculpture of man results in equally well formed his genetic code. The described effect refers not only to the genetic code but also to all the features of man all the time he stays in the niche of mankind.

As the successive niche of man we will separate his immediate surroundings in everyday meaning of the word, i.e. collections resulting from the interpretation of canvas. Those will be other people, permanent residence, and objects. People of similar inclinations create niches corresponding to them. Membership in a group of particular interests or opinions is the membership of the definite niche. This does not deal with any official declaration of membership but the information connection. Any strict connection between people permitting them to be separated as a group is simultaneously a niche. The intensity of the information interchange decides the separation of the niche. Therefore, not always a membership in the definite niche will be the conscious membership. Also not always the membership will agree with desires and ideas of the definite man.

We will next separate a niche of the region in which man lives, the niche of his national status, religious views etc. In turn, we will separate the niche of mankind, the common niche of all the people together with their surroundings. Further, we will separate the niche of our planet and all the entities associated with it, the niche of the universe known to us, etc.

As a matter of fact, the niches, which we have mentioned here, can be formed by complete collections, whereas the conscious perception of man and also the perception of instruments built by him do not include this completeness but merely its fragment represented within the framework of the interpretation of canvas. We should remember this when considering the problem of niches.

The real, reality, man's surroundings

Our interpretation of the real has been formed in such a way that <u>we regard as real what belongs to our common time sequence</u>. Collections included at present in the sequence and in consequence rapidly loosing their information potential are interpreted as reality occurring in the present. Collections, which have been included in the sequence earlier, are the memory of reality.

This manner of interpretation of reality does not mean that we cannot change it.

The surroundings observed by collections are formed as the common interpretation in the given niche. The same thing refers to man as a collection. The surroundings, which we observe, are the common interpretation of people and all the other collections taking part in the situation. A contribution of people to this interpretation can but does not have to be dominant, since they are not the only collections participating in it.

We will often use the determination 'the common time sequence of people'. However, expressing it more precisely, more than one sequence can exist. Every subordinate niche, associated with a group of people participating in the common situation has its own sequence. Man, from his point of view, acting in isolation from other people, also represents a certain time sequence. Niches and their time sequences in which people take part are connected with one another with the intense information interchange. For that reason, situations observed in them are often represented by the very same collections. Their interpretation is similar in different niches but it is never identical. A strong connection between niches in which the canvas is known denotes their high accordance with regard to this interpretation and thus, it offers facilities for synchronizing clocks and for synchronizing their position in the physical space.

This part of the individual sequence of man which is not common to other people, i.e. does not belong to the sequence of the common niche is interpreted as thoughts, imagination, dreams, etc. We customarily treat imagination as the unreal world but the situations in which we participate in imagination and in our common reality are different from each other solely in their range and in the intensity of information exchange. Let us notice that if the intensity of dream or imagination is comparable with everyday reality, a solitary man will not be able to find whether he occurs in the individual reality or in the common one without appealing to opinions of other people. This happens, since the only discriminating factor of our reality, apart from the intensity of information, is its sharing with other people.

Situations experienced in every kind of the time sequence, also those, which we call imagined ones, are important for the formation of the potentials areas and the whole of a collection. The importance is proportional to the intensity of the information interchange accompanying the given time sequence.

The common sequence of people is a more intense variant of the individual sequence of man and by comparing its intensity with sequences of other niches of Area K we not necessarily find that it is most intense. The difference in the intensity between sequences of other niches and our common one can exceed the difference between the common and individual sequences of man. Which of such sequences will we call the reality then? Obviously this one which is more intense in the sense of information. Reality represented by such a sequence can be intense enough so our everyday one will move to the role of imagination or below. However, the comparatively active participation in such a reality requires a proper degree of development from the collection. A weak collection would not be able to participate in intense reality through a long time. In case of a strongly extended collection, its individual time sequence can have the greater intensity than the common sequence known to people representing their reality.

Every man developing his collection and learning the freedom of making use of interpretations is able to form a time sequence of any intensity and contents. For these collections for which such a sequence is common and the most intense one of those observed it means *reality*.

Conclusion. Reality is the local interpretation limited to the definite niche. A collection can simultaneously participate in many realities (time sequences), the examples are imagination and everyday reality. Reality, which is connected with the greatest intensity of information and, at the same time, is common for the collections similar to the observer will be recognized as the basic one by him.

Imagination, thought

Imagination is a collection. Imagining something by oneself is the inclusion of a suitable collection in its own individual time sequence but excluding the sequence of the niche common to other people. The intensity of the information interchange with a collection can increase. Then, if the act of inclusion of the collection in a sequence reaches some of common niches, we will call such an act not imagination but reality, i.e. a situation observed in like manner by different collections of the niche. Imagining can occur in two ways:

1) Reinforcement of imagining dominates over the loss of the information potential (the loss of potential accompanies the inclusion of imagining into the individual time sequence).
As a result of reinforcement, imagining can occur in the potentials area of the niche or if it already occurs there, its position will be reinforced. Including it in the common sequence of the niche will be more probable.

2) A loss of the information potential of imagining dominates accompanying its inclusion in the individual sequence of man.
As a result of this, the probability of further appearance of imagining in the individual sequence of man will gradually lessen. The potential of imagining with regard to the common niche can only be slightly attenuated as a result of such an action. A possibility of the appearance of the situation in the common sequence can also be insignificantly weakened. The probability of the appearance of the situation represented by imagining with the participation of man whom we describe can lessen to a far larger degree. However, this also depends on the thread of time associated with the imagined situation. The sculpture of the niche can force the participation of persons of a definite kind in the situation but it never forces the presence of concrete individualities. For that reason, it is easier to change an individual collection (persons) taking part in the situation than to prevent the inclusion of this situation in the common time sequence.

Knowledge about these two ways can practically afford possibilities for a conscious influence on forming time sequences and for avoiding an unconscious, undesirable influence on their development. The way described in item one allows for increasing the possibility of the observation of an imagined collection in *reality*, i.e. in the common time sequence. The way described in item two weakens the possibility of participation of the given man in the situation observed in the common time sequence.

In practice, the first way is thinking about the situation developing its particulars. The second way we customarily call *experiencing* the situation in our imagination. This kind of way is associated with the effects in the form of emotions.

Desires, emotions

In the previous chapter we have distinguished to ways, which a thought can adopt.

The first way, being reinforcement of the collection of a thought is interpreted as desire. The second way, receiving information from the collection is interpreted as emotion. The reception of information from the collection does not lead to decreasing its extension but to lessening its information potential with regard to the receiver.

Expressing it more precisely:

Desire is the interpretation of sending information to another collection, i.e. reinforcing it.
Emotion is the interpretation of receiving information, i.e. weakening the information potential of a collection with regard to the collection receiving information.
Desire is not emotion.

Our everyday interpretation of these concepts is associated with the interpretation of time. The above-described kinds of the information interchange refer to the potentials area. Desire is reinforcement of the selected collections of this area, whereas emotion is receiving information from collections included at present in both the individual and common time sequences. We customarily divide emotions into two basic kinds: negative ones as suffering and positive ones as joy. This differentiation results from the comparison of the individual and common time sequences. The information interchange which results in assimilation of these sequences is interpreted as positive emotions. The increasing differences between these sequences lead to the interpretation of negative emotions. Besides, a way of the interpretation of emotions depends on its sculpture, as every interpretation occurring in the niche. If the connection of the given situation with positive emotion is formed in the sculpture, such a tendency will occur in every individual case.

When collections, which till now have been observed and extended only in the individual time sequence are also included in the common sequence, this is understood as the fulfillment of definite desires. When it does not happen, the interpretation of suffering appears and it is greater, the more collections of the individual time sequence have been reinforced and which have not appeared in the common sequence so far. For example, when man in his thoughts, i.e. in his individual sequence, supports (reinforces) the image of himself as healthy, whereas in the common time sequence he appears as ill, suffering is the interpretation of this difference. If the situations observed in both sequences were exchanged, suffering would still be the result of it, since its source is unfulfilled desires.

The character of the situation observed in the sequences is not essential for arising the interpretation of joy or suffering. The dissonance between the individual and common time sequences is merely essential. However, if the connection between the given situation and feeling of the definite emotion is formed in the sculpture of the niche, this can decide the way of the interpretation of emotion.

Let us arrange the described processes more accurately:

1) Man begins to reinforce a selected collection.
2) The collection is included in the individual time sequence of man.
   It has a comparatively low information potential in relation to the potentials area of the common niche. Therefore, it is not included in the common sequence.
   The common time sequence is interpreted as *reality* for people taking part in it. Since the collection about which we are talking does not occur in this sequence but only in the sequence of one of people, it is interpreted as his thought.
3) Reinforcement of a collection (thought) is interpreted as desire.
   The mentioned thought represents a certain situation, which is observed at this moment as imagining only by the given man. Reinforcing and developing this thought is a desire to observe the situation in a more intense way. Desire is not emotion. If emotion appears on this stage, this implies that the information has been sent from the reinforced thought to man. Practically this situation is not to be avoided, since reinforcement of the collection is ordinarily accompanied by its observation. Emotions can influence the development of new or modification of the previous desires.
4) Participation in the common time sequence is associated with the intense information interchange with the collections already included in it. Reception of the information is accompanied by the interpretation of emotions. The way of this interpretation, i.e. receiving emotions as negative or positive ones, like every interpretation depends on the sculpture of the niche in which it is performed. If the connection of the given situation with a positive emotion

is formed in the sculpture, such an interpretation will be used in the most individual cases. Apart from the sculpture, a significant influence upon the interpretation of emotion is exerted by the comparison of information flowing from the common time sequence to the situation in man's own individual sequence. The collection reinforced by man has not appeared in the common sequence yet. This can be the reason why the flowing information from there will be interpreted as a negative emotion.
5) The further development of the situation depends on the way of using thoughts which man will apply (the two ways are described in the previous chapter). If he still reinforces the mentioned collection, this can be the reason of arising negative emotion accompanying the participation of man in the common time sequence. After some time, when a collection at last appears in the common sequence, a loss of the information potential will be accompanied by a strong positive emotion felt by man. However, if man chooses the second way of thinking founded on the previous discharging of the potential, it will anticipate the appearance of the collection in the common sequence. The emotional effect will be similar but of weaker intensity. In other words, man will experience the given situation in his imagination instead of in reality, i.e. in the individual sequence instead of in the common one.

Summing up, desires and emotions are interpretations of the lower level than the interpretation of time. If the interpretation of differentiation is known, i.e., a collection discerns between its individuality and external collections, the direction of sending information can be distinguished. At this moment the convenient conditions appear for the interpretations of desires and emotions. Let us notice that these are not yet exactly such desires and emotions, which we know from our everyday life, since there is no interpretation of time yet. On this level, desires and emotions are <u>the disconnected</u> individual acts of identity. Therefore, one cannot talk about their effects. Let us emphasize once more that these kinds of desires and emotions are not such as we know in our everyday life.

The situation alters when the interpretation of time sequences appears. The time connections appear in the form of the threads of time also dealing with emotions. Time sequences and differences between them appear. Since these interpretations occur on a higher level than desires and emotions, every collection, which is subject to the interpretation of time, also knows the interpretation of desires and emotions.

Particular interpretations resulting from differences between time sequences appear. They are revealed in the form permitting such kinds of desires and emotions as positive and negative ones to be distinguished. Such emotions as joy and suffering appear on this stage. Suffering is a special kind of interpretation, since it results from the dissonance between individual and common time sequences. Since the existence of these sequences is founded on the difference between them, this constitutes an unceasing source of the interpretation of suffering. Every desire modifies the individual time sequence and brings about the growth of the mentioned difference between sequences. Thus, at the same time, any desire causes the desiring collection to feel suffering. However, not always one can observe the immediate time correlation of these events. Suffering is not a transitory feeling specific only for people as we often understand it but it is known to every collection which is subject to the interpretation of time. As long as a collection is subject to this interpretation, it can only decease the feeling of suffering. It is completely missing as it resigns from identifying itself with the definite time sequence. This does not have to mean that it utterly resigns from this interpretation, such freedom of identifying itself as has been described in the chapter 'Timelessness' is enough for it.

Every desire is reinforcement of an adequate collection representing the aim of a desire. Such every action changes the state of the collection at least to a small degree and also the whole of a niche (to a comparatively small extent). Therefore, <u>there are no desires, which could not have any effects</u>. In case of the extended collection such as man, not every desire is consciously controlled, since the area of the physical consciousness of man does not cover the whole of his collection. But every desire has an effect. The connected effects of desires of all the collections of the niche influence which collections and threads of time will be included in the sequence of the niche, i.e. they influence the way the common *reality* of the collections belonging to the niche will be formed. The collections reinforced as a result of a desire run a good chance to be included in the time sequence of the niche but it does not mean that they will be included at once. The immediate effect of a desire is the change of the potentials area. On the other hand, the effect in the form of a variation in the time sequence is not prompt and it can be observed with a considerable delay. The appearance of the collection to which a desire refers in the common time sequence fulfils the desire. Whether it will be easy or difficult to fulfil depends on the quantity of information, which must be sent to include the objective of a desire in the sequence. <u>The weaker an extended niche and its sequence are, the easier the desire is fulfilled</u>. In case of man, it will be easier to fulfil a desire, which deals with a small niche, e.g. several persons than such one which refers to millions of persons. It is easier to do a thing, which is not consistent with the sculpture of the niche in a circle of few people than in front of crowds of people. The reasons of this lie in the quantity of information (a number of the acts of identity) which is used in time sequences of different groups of people.

Particular desires and emotions in themselves are collections. They can be reinforced and sent to other collections. If they are sufficiently extended, they can display distinct features of the collections, which have supplied them. They can have their own desires. Inasmuch as they have been reinforced, they can act consciously and their action does not have to be consistent with the primary intensions of the collections supplying them. If they are so extended that one can speak about their conscious action, this denotes that it is possible to communicate with them like with every collection of this kind.

Emotion, which a given man feels did not necessarily have to originate owing to him. It could be sent to him by another collection. When man began to reinforce the identifier of emotion, it is probable that it was identified with one of the representatives of the sculpture of this emotion. In this manner, he obtained the identifier of emotion, which is already extended within the niche.

As we have mentioned, desires and emotions belong to a comparatively low level of interpretation. They exist on a much lower level than brain and human body. Excluding the lowest level of interpretation, desires and emotions are universal phenomena amongst collections and they do not depend on the fact that whether the interpretation of canvas, i.e. a physical body or its equivalent has been formed in the given niche. The fact that these phenomena are universal does not mean that in case of other collections, their intensity is such as with man. It will be much less for the collections less extended and from our point of view it will be neglected.

A desire contributes to the increase in the difference between individual and common time sequences. The more intense they are, the stronger the interpretation of separate time sequences is. The clear distinction of the observer's own individual sequence from others signifies a stronger feeling of the observer's own individuality. Consequently, the conscious actions of the collection (if it is a well extended collection) will be strongly oriented to its own time sequence. In other words, such a collection will be to a larger degree engaged in its own time sequence and it will be much more oriented to its own individual reality, which is its sequence. If we take man, for example, we can say that the intensity of man's ego is the intensity of identifying himself with a concrete time sequence and distinguishing it from others. As one can hence conclude, desires reinforce ego.

On account of the similarity of identifiers and the existence of the sculpture, a collection appearing as its fulfilment in the sequence of the niche does not have to be the very same individual collection with which the desire dealt. The degree to which fulfilment will resemble the desire depends on the degree to which the individual collection of the objective of the desire will be extended and the precision of its development, i.e. how it will differ from the threads of time formed in the sculpture of the niche. When the desiring collection reinforces the collection of the objective of desire, not determining precisely the role of individual collections (including its own role), fulfilling such a desire can refer to quite other collections than those which participated in the desire. A similar effect will take place when the very same situation is the subject of desire of many different collections, since each of these collections will reinforce the objective in a different way.

The simultaneous existence of desires, which we interpret as the opposed, is reinforcement of two different collections being the objectives of desires. Therefore, this does not cancel both desires. Adequately reinforced, they both will be fulfilled but not at the same time or not in the very same time sequence. A similar situation will take place when a desire is suppressed and the contrary one appears in its place. The contrary desire will not restrain the previous one from fulfilling and, if it is stronger, the final effect can agree with the intensions of the desiring collection. The real recall of a desire cannot rely on reinforcement of any collection. This can be achieved only by weakening the collection of the objective of desire, i.e. by decreasing the information interchange with it, in other words, by forgetting it. In practice, it can be difficult to do, since the majority of collections possessing the identifier of the objective of desire would have to perform this action.

We have stated earlier that every desire is accompanied by its effect. This deals with both fulfilled desires and those yet unfulfilled. In the first case, the effect is more distinct, since fulfilment is accompanied by the more intense information interchange. Fulfilment relies on the inclusion of the collection representing a desire in the common time sequence. Thus, the intense information interchange refers not only to the collections, which bore the greatest part in reinforcing the desire but also to all those participating in the given sequence. This is the reason why a reaction to fulfil a desire can be stronger than the desire itself. By strength I understand the intensity of the information interchange. The reaction to a desire does not have to agree with the intensions of the collection from which the desire derives. In case of the contrary desires, this can bring about the oscillation, i.e. alternate fulfilment of such desires.

The desiring collection reinforces the collection of the objective of desire extending its identifier. By doing this, it sends its identifier to the objective of desire and its component collections. At the same time, it reinforces its own identifier being in possession of these collections. In this manner, <u>a desire connects the desiring collection with that of the objective of desire</u>. By connecting, I understand their mutual information interchange. Thus, the possibility of the simultaneous inclusion of both collections as linked together in the common time sequence will occur. If the situation illustrating the objective of desire anticipates it, the probability of its fulfilment increases. If, in the situation being the objective of desire, the desiring collection does not occur, we can finally observe the desire fulfilled twice, once in the way consistent with imagining of the desiring collection, i.e. without its participation. For the second time it is fulfilled in like manner but with the participation of the desiring collection. The second time results from a strong connection of the desiring collection with its objective, which can cause reversal of roles of both collections. The described effect is quite independent of the kind of desire. Such concepts as 'good' or 'bad' desire do not occur yet on this level of interpretation.

Morality, the good

The difficulties, which we have with the definition of these concepts, result from the high level of interpretation on which we attempt to do it.

The interpretations connected with the time sequence, i.e. desires and emotions decide what is good and what is not on this level. The divergences between the individual time sequences and those of the niches of social groups will account for the divergent defining of morality. They will also account for dividing events into favourable or unfavourable ones, i.e. into consistent or inconsistent with its desires. This refers both to the individual collections of people and to those of social groups. This, in turn, leads to the formation of bipolar ethics and morality founded on the division of phenomena into desirable and undesirable ones.

Since the interpretation of canvas is the most characteristic feature of our common niche, its properties have the greatest part in forming the interpretation of morality. We interpret these characteristics of a collection as physical objects including biological bodies.

Descending to more basic levels of interpretation, below the interpretation of desires connected with time sequences, we will notice that their degree of extension is the fundamental difference between collections. The stronger the collection is, the more conscious its action is. On this level <u>the good denotes such a kind of action, which reinforces the collection</u>, i.e. every act of identity dealing with it. Since the contrary of the operation of identity does not exist, <u>the good also does not possess its contrary on the basic levels of interpretation</u>. No collection is able to 'do harm' to another one, but it can only more or less reinforce it. Whereas the collection itself, if it is sufficiently extended, it consciously chooses the interpretation which it will apply to others.

The contrary of the good arises on the high levels of interpretation as resigning from a portion of generally accessible information.

Responsibility

In the previous chapter we have stated no collection is able to do harm to another one. The statement refers to the low levels of interpretation. As long as we are not able to stay on these levels, we cannot offhand use this law to our present state. We lose much information circulating around us, when we stay on the level of interpretation of canvas. The layers of our complex interpretations act as the filters of information. Let us not neglect them until we do not get rid of them.

On our everyday level of interpretation, we must consider that every action will be judged as positive or negative by our environment and by us. What does the concept 'judgement' denote? This is the connection of emotion and desire known to us from the previous chapters. The observation of the event is the reception of information from the collection included in the time sequence. The reception of information is interpreted as emotion. As a result of the secondary interpretation, emotion on our level is recognized as positive or negative according to whether the difference between the common time sequence and individual one increases or decreases. This process has been explained in the chapter 'Desires, emotions'. The reaction of collections to their own emotions follows then. This reinforces the selected collections in order to bring about the agreement of the common sequence with their own one. The reinforcement is nothing else but desires. Each of them exerts an immediate influence on the potentials areas of the mentioned collection and all the niches with which it is connected. Thereby it affects the further formation of time sequences connected with these niches.

Summing up, emotion resulting from the observation of events has brought about the change in the future formation of time sequences. This process is illustrated by the concept of 'judgement'. This does not express merely passive verbal judgements, as we normally understand it. Alternate emotions and desires result from the way in which the interpretations of the potentials areas and of time sequences are formed. They require unceasing reinforcement of the collections from the potentials areas and discharge of their information potential in the time sequence. The described process does not have to be named judgement but simply a reaction to the events observed in the sequence.

We should take account of a multiple reaction to our every action when we stay on this level of interpretation. Our actions are founded on reinforcing the definite collections. This is its first stage. Next stages constitute a reaction to the situations included in time sequences as a result of the actions from the previous stage. These can be individual or common sequences. The observed distance of time may vary and depend on the intensity of supplying the collection from the potentials areas and on the degree of extension of the whole of these areas.

The reaction relies on reinforcing the collections of situations, which most probably will contain identifiers of the collection, which has begun the chain reaction. In this manner, the reaction will form the future events in which the mentioned collection will take part. From everyday point of view, this denotes that our actions affect the shape of future events in which we will participate. This seems obvious, but let us notice, that it concerns all kinds of actions in every time sequence with which we are connected. This includes both thoughts, imaginings and the smallest common time sequences, i.e., the situations when man acts solitarily without the knowledge and presence of other people in our reality.

By reason of ambiguity of identifiers, the reaction will not always deal with the collection which has started it although, according to the reacting collections, it should be so. This means that other collections belonging to one of the common niches can bear consequences of the action of the given collection. This effect can be named as bearing the group consequences or as group responsibility. In case of people, the act of one man can result in the reaction directed to man himself or to other people in one of the niches to which that man belongs. Which niche it will be depends on the kind of the reaction in which the mentioned man will be interpreted as a member of definite niches. For example, a painter handing over some flowers to a certain person will for the most part be identified as a concrete person but also as a painter to a certain extent. Thus, the consequences of this action will be distributed between an individual person and a niche of painters. This does not denote that the whole of the niche will suffer from the effects of this action. This can deal with only one or with a few of its representatives. From our point of view, these will be some casual representatives of the painters' niche. Distribution of the reaction to many collections will cause its less intensity observed in each of the individual cases.

As a result of the action, the collections from the potentials areas are not only reinforced to be included in time sequences. Every action will develop the sculpture of every niche in which it is observed. Especially, the more often the action is repeated and the more intense information interchange accompanies it, the stronger the sculpture will be. Every action increases the probability of its repetition in the future. In case of man, the influence of events on the sculpture of his individual niche will be interpreted as forming his character. In case of social groups, we observe the changes in their sculpture as the variations in repeating behaviours of their members. Every action within a niche characterized by the interpretation of canvas, brings about the change of laws, which we call the laws of nature. However, because of the strong development of the niche, such a change can be observed after a long time.

An additional effect of the action (reinforcement of other collections) is a stronger connection between the acting collection and all those whose identifiers occur in the situation being reinforced. This deals with any kind of action, also the one that is directed 'against' a certain collection. For a certain period of time, such an action can be interpreted as one bringing the prospective results but, later on, it can result in the connection of both 'adversaries' to one common situation. The more probable it is, the more intense the 'adversaries' or 'followers' the collections were, since both of the cases lead to the common exchange of identifiers and to reinforcement of the situation in which both collections participate. It will be easier to observe the effect, if at least one of the mentioned collections is comparatively well extended.

Acting in the area of thoughts, we should pay particular attention to the role of negating concepts. As we know, the operation of identity cannot be diverted. It is not possible to negate the identity. Expressing our desire verbally, we should not use negation, since the collection occurring in desire will be reinforced but not its contrary. For example, a desire expressed as 'I will not grow fat' will reinforce all the concepts occurring in it and also the concept of fatness but it will not reinforce its contrary, which we conjecturally assume. The contrary is the effect of our additional interpretation of the mentioned statement. When it is correctly expressed, it should state 'I am slim'. On the other hand, it would be better to forget the word 'fat' but not to avoid it, since <u>conscious avoiding of a given collection denotes its</u>

reinforcing. The fact of avoiding forces attention to be focused on the avoided object and, as a result of this, the information interchange with it.

<u>Negation facilitates logical reasoning for us but it is inapplicable on the level of the interpretation of desires</u>.

Conscious formation of threads of time

The situations, interpreted by us as the real world, are grouped in the threads of time. They can be liberated by time keys. For these reasons, the phenomena are inclined to be repeated in the definite sequence, which gives the impression of the established order amongst the phenomena.

The threads of time formed in the sculpture of a niche derive from all the acts of identity, which took place in the niche at any time. Its sculpture is subject to the continual modification. This enables new threads to be consciously formed and to model the existing ones. The threads can be used as keys serving to obtain the definite phenomenon. The degree of the extension of the niche and the possibility of the collections forming the thread determine how much information is necessary to send in order to form a new thread. In other words, this is determined by a number of collections, the degree of their extension and the time which they devote to reinforce of the collection of the thread. The change of its own individual sculpture does not have to be easier than the alteration of the sculpture of a more extended niche brought about by a larger number of collections. The individual sculpture of man, as we have mentioned in one of the previous chapters, includes his character. For an individual collection, his change can be equally difficult as alteration of the sculpture of the niche common for people and made by the whole society.

Habits

These are the repeated behaviours of man resulting from the sculpture of his individual niche and from the sculptures of all the other niches in which a given man stays. The threads of time formed in the sculptures act as a matrix, giving a determined interpretation to the collections included in the sequence. We can observe the action of the sculpture in relation to man under the form of habits and addictions.

Man finds himself in a hierarchy of niches and the sculpture of each of them influences his behaviour. A group of people exchanging information between themselves more intensely will be inclined to make their habits alike. Only a part of information is contained in the consciousness of man. Its real quantity and character can remain unconscious. Since the environment of people is formed as the common interpretation of many collections, one can estimate that the quantity of information transferred beyond everyday consciousness of man is much larger than that of which he becomes conscious.

The action deprived of habits requires the independent formation of collections representing the situations precise enough to lessen the influence of the threads of time contained in the sculpture to the last degree. Thus, it requires a comparatively large quantity of information. The action in which habits prevail does not require such an intense exchange. The weaker information interchange of the collection with the niche denotes a weaker connection between them. The increasing succumbing to habits, i.e. the weakening connection with the niche can mean an approaching loss of the connection between man and our common niche. To a certain degree this can also be the cause of his abandoning the given niche.

Contacts of man with other collections

Only one way of communication of collections exists: the act of identity. When collections, interpreted as people, communicate between themselves we can see people talking with each other, writing, making gestures. This way of interpretation has been formed in our niche as a fragment of its sculpture and it decides how we perceive the acts of identity performed between us. The sculpture of the niche changes continually, a way of the interpretation of the communication between people was different in the past than it is now and it will be different in the future.

When a collection of man exchanges information with a collection interpreted as a chair, the sculpture of our niche forces a picture of a man holding a chair or sitting on it. The main stream of information associated with this event is interpreted in this manner. As we know, the interpretation relies on the loss of a portion of information. Therefore, the weaker acts of identity accompanying the situation will not be included in the time sequence of man. The information interchange occurring beyond the conscious action may refer to every kind of component collections, also those not resulting from the interpretation of canvas such as thoughts. Let us be engaged with the collections representing the information exchanged in an unconscious way. These collections which are derivatives of the interpretation of canvas, can be detected by our instruments, which also result from this interpretation. These are such objects as finger-marks, sweat left on a chair, dust, bacteria, etc. A chair is a collection much less extended not only than man but also than majority of man's component collections. Therefore, its every contact with man can leave clear traces in the collection of this kind and not only in the form of collections resulting from the interpretation of canvas. Man meeting <u>any</u> collection will leave or extend his identifier in it, composed of the identifiers of some of his component collections. The current state of the collections participating in the contact and the threads of time forming the contact determine which identifiers will be included. Apart from his finger-marks, man will also leave his identifier reinforced by the contact, i.e. a part of his thoughts as well.

Finally, it is possible to exchange information between people who personally (in a conscious way in the interpretation of canvas) are never in touch with one another. The role of a carrier of the information interchange can be played by any object with which people are in contact not necessarily at the same time. On account of a little intensity of the exchanged information, it will rather not be observed in time sequences, i.e. it will not be perceived by people in a conscious way.

By the contact of man with another collection, one should understand not only his physical contact in ordinary meaning. The contact does not have to deal only with collections which are subject to the interpretation of canvas, i.e. physical objects but all the others such as thoughts or collections of which man is not aware. Every thought observed by man in his individual time sequence denotes the information interchange with it and with its component collections. If such a thought represents a situation in which another man participates, this man will also become the aim of the information interchange at this moment. If a thought is feebly reinforced, the information interchange with that man will be also insignificant. It can be too weak to be included in the individual sequence of the other man. Otherwise, it will be included and observed by him as his thought. As a result of a strong thought of one man a similar one can be then observed by another man. If that man fails to read identifiers in it of the man who has begun to reinforce the thought, it will be interpreted as his own thought. Since every man has identifiers of other people, his thoughts will be transferred to them even without his conscious effort. It is only up to them whether they manage to include such a thought in their individual time sequence. The inclusion of a thought in a sequence permits its existence to be found. In order to find its source, man must be able to read the identifiers contained in it.

Scale of information concentration

The information accumulated in a collection is a feature permitting us to foresee how the collection will behave in relation to the others. A comparatively large accumulation of information in an individual collection denotes a large number of the acts of identity and, at the same time, a strong feeling of existence and extended consciousness. This not necessarily signifies a strong feeling of one's own individual existence. This depends on the relation of the intensity of the information interchange inside and outside of a collection, i.e. on the distinction between the inside and the outside.

Using the assessment of the information concentration, we can range the collections known to us according to the following scale:

1) Elementary particles.
2) Chemical compounds, objects that we call 'inanimate'.
3) Simple organisms.
4) Plants.
5) Animals.
6) Man.
7) Supercollections.
... etc.

The division is approximate and concerns merely the common niche known to us, people. It takes into account the individual concentration and not that of the sculpture of a collection. Man is not placed as the last item, since there is no foundation to assume that he is a collection of the possibly maximum information concentration (let us remember Principle 2). Collections of the individual information concentration greater than the average of man we will conventionally name ***supercollections***. In the presented diagram, supercollections are represented by item seven and all the subsequent ones. The boundaries between items should not be treated sharply, for they are approximate and they can penetrate one another. With the exception of supercollection, the division is exclusively founded on the observation of the features of canvas and it does not comprise the whole of the collection what makes it still more approximate.

Considering the described items, we can expect both the increase in gathering the information and in the conscious control over it. However, one should not expect to appear or develop such particular ways of thinking as logic. Such ways of thinking can be formed in a given niche but they are not indispensable for the action of a collection on any level of the information concentration.

The items presented above, first and foremost, cover the collections observed as a result of the interpretation of canvas, i.e. possessing a definite position, dimension and mass in the space-time known to us. The canvas is merely one of the features that can be revealed by collections and only some of them have this feature. For example, thoughts are collections, which do not have the features of canvas. They have not been included in the items mentioned above, since the degree of their concentration can vary much and it can cover all the items.

Together with the increase in the control over the information interchange, the control over the interpretations also increases. This control can achieve such a level where the existence and the kind of the interpretation of canvas will be the conscious choice of the collection (the supercollection from our point of view).

Supercollections supplied by people

In the chapter 'Contacts of man with other collections' we have found that directing attention (thought) to another collection results in sending the information to its identifier.

Let us assume that a collection exists, which a large number of people pay attention to. The more people take part in it, the more information will be sent to this collection. The quantity of the transferred information will also be proportional to the intensity of the contact and to the duration of this situation.

If many people are busy for a long time, it may happen that the information gathered in the collection will reach a level considerably exceeding the level of an individual man.

What features will such a collection have?

1) The features with which the collection has been supplied will prevail or, in other words, those will prevail which people attribute to this collection.
2) Apart from the features consciously attributed (the previous item), the collection will also acquire the characteristics of the people supplying it have and characteristics imparted in an unconscious way.
3) The information concentration considerably greater than that with a common man also denotes a greater opportunity for the conscious action of the collection. For this reason, in the contacts with an individual man and a supercollection, the latter will decide the way of the contact and also the interpretation of the way the supercollection will be perceived.

A natural means of communication of man with a supercollection is a thought.

4) Since a collection is supplied with thoughts of people, it will probably maintain such a situation. This signifies its positive attitude towards people and its positive reaction to their needs. This would be a kind of symbiosis between them and the supercollection. However, because of the reason mentioned in the previous item, the motives for the action of the supercollection can be difficult for man to examine and estimate.

In practice, supplying a supercollection and making use of its aid seem to be reasonable if the way, in which the action is performed, is understood. A group of people being in the symbiosis with a supercollection will manage much better in the critical situations than a group of people without such an aid. Every thought directed at the supercollection is a thought supplying it. It is not important whether in our interpretation the thought is 'positive' or 'negative'. The only way to weaken a supercollection is to limit its reinforcement. In practice, this denotes forgetting it. This does not bring about the annihilation of the supercollection but it can break off its connections with our niche.

From our point of view, the existence of supercollections with positive and negative feature is possible. This depends on the kind of thoughts supplying the supercollection.

We have described supercollections supplied by people. These are supercollections of a special kind, not every one of them must be of such origin.

Supercollections supplied by other collections than human

Since man is not the only collection able to think, the existence of supercollections similar to those described in the previous chapter and supplied with other than man's thoughts is possible.

Such a creation will have all the features of a supercollection. It will not display human features but those characteristic for supplying collections.

Such collections as plants, animals or man will not supply the very same supercollections on account of too great differences between their niches. The supercollection will be connected with a definite group of supplying collections similar to one another. Collections such as animals, plants, minerals, thoughts, other supercollections, etc, can be the supplying collections.

Can we notice such supercollections in practice? These collections have been reinforced by thoughts, i.e. by collections, which are not subject to the interpretation of canvas. For this reason, the supercollections arisen in such a way can be inclined to remain beyond this kind of interpretation, i.e. beyond our everyday perception founded on interpretation of canvas. However, there are no obstacles for coming into contact with them by means of thoughts.

Free will, active action

The qualification 'free will' is rather a dim concept understood in different ways by different people. However, let us retain it for the sake of its importance for us.

The problem of free will is very similar to that of the cause of action (the chapter 'Cause of action').
Man is a set of many different collections. Each of them, like him, has its consciousness and goals realized by the acts of identity. Man is a composition of these consciousness and goals. It happens that one of the component collections forces its will upon others.
This can be tested experimentally. For this purpose, please try to think only and exclusively about one object for five minutes. After this time, please estimate how many undesired thoughts appeared during this experiment. A common man is not able to prevent the appearance of miscellaneous thoughts even for a short period of time.
These undesired thoughts are collections which like man, although to a lesser degree, have consciousness and they would

desire to have free will. These collections exchange information with man. They can be strongly connected with him through the extended identifiers but they are not his exclusive 'property' and they live their own existence. This refers to all the collections of which man is composed and also to his thoughts. Moreover, these collections are in a similar situation in relation to their own component ones. The described situation does not deal with man only.

Everything what man does and thinks of is a composition of the influence of a great number of collections, a whole chain of collections connected directly and indirectly with one another. Looking from this point of view, every thought and every action of man have their source in the action of other collections. In reality, the action of many other collections is that what man registers as his own action.

All the other collections are in the very same situation. The action of each of them reveals the action of the others. We can come to the conclusion that no collection really decides its action. That brings to mind our considerations about the existence of a collection. No collection can exist independently. The existence of each of them is founded on the existence of its identifiers in other collections. A similar situation is with its action. A collection cannot act of itself. Its action lies in the action of other collections. This does not imply that the collection does not act. Its action is <u>the interpretation</u> of the arisen situation. The interpretation is of the local character and its scope is limited by the niche, which time sequence the collection regards as its reality. From the viewpoint of the collection, both the existence and its free will of action are unshakable. Looking from the global point of view, no collection acts independently.

However, let us remember that 'the global point of view' is also only the local interpretation. The collection, which has identified itself with Area K, i.e. finding itself beyond the range of our concepts, has a real global view. From that and from the point of view of Area B the situation can be quite different.

Moving in the world of collections we are not able to fully analyse this situation but we should not forget that it is possible to get to know Area B and to look from its point of view.

What is the answer to the question about the existence of free will?

The answer is simple in its sense but it is difficult to formulate it verbally. The answer is relative.

<u>As long as you are a collection, you can think that you know what free will is and that you have it in the very same manner in which you think that you exist</u>. Remember that your concepts of existence and free will are only successive collections and do not treat them as ultimate and invariable. If you try to look globally, it will turn out that no collection acts of itself but the action of other collections is only reflected in it.

Area B not being a collection is a foundation of the existence of this situation.

Destiny?

Destiny is customarily understood as the determined fate forced with some time advance. From the point of view of our knowledge about collections, one can understand it as finding some properties of time sequences. The strongest collections from the potentials area of the common niche included in its time sequence are the events observed as the reality currently occurring. Their information potential with regard to a given niche decides what collections and in what succession will be included in the sequence. Also the sculpture of the niche decides the succession in which they will be observed. Whereas, the information potential results from all the early operations performed on the collections from the potentials area. The mentioned operations reinforcing the selected collections occur beyond the common time sequence, so they are not observed as *reality* in our everyday understanding. However, they can be observed as thoughts and concepts in the time sequences of other niches. If the information interchange of individual collections is less than the degree of development of their common niche, from the viewpoint of the members of the niche, the process of reinforcing the selected collections from the potentials area can last for a long time.

Let us sum up our knowledge about the succession of events:

1) The causes of the observed events do not lie in the situations directly preceding them The succession of events results from the threads of time formed in the sculpture of the niche. There is no direct cause-and-effect relationship between events. The statement that such relationships exist results from the interpretation of recurring situations connected with the loss of a portion of information exchanged between collections.

2) If we talk about the cause-and-effect relationship, it refers to the events observed not immediately one after another. This results from the cycle of the successive including collections in the time sequence. A certain event is being included and as a result of this, adequate collections from the potentials area are reinforced, which will be included in time sequence much later in time (as a result of much earlier events). <u>Reinforcing the definite collections is an action separating events in time and simultaneously connecting them in one interpretation of cause and effect</u>. This interpretation belongs to a lower level than that known to us from our everyday life and for which the succession of events forced by the sculpture of the niche is the basis connecting cause with effect. In the previous item, we have found that events immediately following one another are not connected with casual relationships. This results from the fact that a certain period of time

devoted to reinforcing the selected collections from the potentials area must occur between cause and effect.

3) If the selected collections from the potentials area are strongly extended, one can assume with great probability that they will be included in the time sequence. By analogy, there is little probability that a feebly extended collection will be quickly included in the time sequence. The possible actions of a collection to prevent or to accelerate a definite situation in the time sequence can be inadequate on account of the limited information potential of the collection. In such a situation, we can talk about 'destiny'. However, no event is ever foredoomed, since there is always a definite quantity of information, which can change it. In this case, it is important that the collection desiring to make some changes has the opportunity for doing so. The more the given situation is developed as a collection from the potentials area, the most difficult it is to alter it.

In such a developed niche in which people move, reinforcing the situations can last very long from their point of view before they appear in the common time sequence, i.e. in our reality. However, if a situation has been reinforced for a long time or by a greater number of people, it will be difficult to prevent it from being included in our reality. We can name the situation 'destiny' as something that it is difficult for us to change. However, let us remember that we ourselves created the situations before, in which we are participating now. On the other hand, what we are doing at present, including our actions and thoughts, determines the situations, which we will meet in the future, also in the further one. This is not a metaphor, the situations which we create by means of our thoughts are of the same kind as those, which we regard as real. The difference is such that they are included in different time sequences. We call 'thought' what we observe in the individual time sequence and we name 'reality' what we observe in the common time sequence. Developing the time sequence is determined by the potentials area of the niche. Desires are the reinforcing of the selected collections from this area (they are described in the chapter 'Desires, emotions'). Desires are the causes of developing the potentials area and future situations in which we will participate. Here we do not deal with desires, which we openly declare, but with these, which we really feel. The niche the time sequence of which we interpret as reality does not belong to people only. Not only the desires of people develop their time sequence. Lesser or greater influence is exerted by all the members of the niche, animals, plants, our planet as a whole and also by other collections associated with the niche whose existence we are not aware of.

Example with a man and a vase

In order to better understand the principles of the operation of a collection, I will describe an everyday life situation.

The situation seen with the eyes of a man is described in italics. Normal type is used for the Secral's interpretation of the situation. The description from the viewpoint of the high level of interpretation is presented in italics, whereas that from the viewpoint of the low level of interpretation is given in normal type.

A man can see a red vase standing on the table.

> One of the component collections of a vase is the collection customarily interpreted as red. Although a vase can contain many other collections interpreted as colours, this one is the strongest.
> All the collections, which intensely take part in the situation, form a niche. Collections of the greatest information concentration are of the greatest importance in forming the time sequence of this niche. Every collection taking part in

the situation belongs to miscellaneous niches to a different degree. It can participate in the time sequences of these niches to the very same degree. How strong the connection of these collections with the niches is, decides how strong influence their sculptures will exert on the behaviour of these collections. The identifiers of a vase and the whole of the situation (a vase standing on the table in the definite surroundings) are possessed by all the collections (including a man) which have met this situation and others to which the identifier has been given.

A collection of the situation is included in the man's time sequence. This determines that the man can say that he witnesses and participates in this concrete situation.

A man wants a white vase to stand on the table.
A man reinforces a collection representing another situation illustrating the table with a white vase standing on it.
From now on, we are dealing with two collections representing such situations, one with a red vase and the other with a white vase. The first collection is strong, the second is feebly extended but it is reinforced by a man.

A man more intensely wants a white vase to stand on the table.
A man reinforces the collection in which a vase is white. Reinforcing relies on sending identifiers of the collection being reinforced to other collections known to him also to its own component ones. Reinforcing relies also on direct developing of the possessed identifier of the desired situation.

A white vase is standing on the table.
(A man has taken away a red vase and in its place ha has put a white one or he has repainted the vase).
The collection with a white vase has become stronger than that with a red one. Expressing it more precisely, the

connection of the collection with a white vase and with the niche has become stronger than with the collection with a red vase. As a result of this, at a certain moment the situation with a white vase and not with the red one is included in the time sequence of the niche.

The moment of change is a thread of time. On account of the existing sculpture of the niche, this thread will be inclined to assimilate with one of those already extended and contained in the sculpture. The exchange of a red vase into a white one will be observed in the shape of one of the typical situations developed in the sculpture. In our case, this will be the thread of time illustrating a man bringing a white vase and putting it in place of the red one or a man painting the vase or any other means known to us, i.e. well developed as the thread of time and contained in the sculpture of the niche common for us.

The appearance of the thread of this kind is not necessary to observe the situation of the exchange of colour or for any other situation to appear. However, if the sculpture of the niche is strongly developed, it will force one of its threads, which is the most similar to the arisen situation to appear in the time sequence.

Let us remember, as it has been described in the previous chapters, that one does not deal here with any real exciting force. The effect of exaction arises as a result of the assimilation of identifiers of the collection, i.e. as a result of neglecting a portion of information obtained by the observer.

This effect is utilized by a man when, after expressing his desire to change the colour of a vase, he reinforces one of the collections known to him and which customarily accompanies such a change. In this manner he makes use of one of the threads contained in the sculpture of the niche. If the man resigned from a typical thread and tried to create a completely new one, this would require sending a greater quantity of information, maybe too great for his current possibilities. If a given man has this quantity of information, he can create new threads unknown in the sculpture of the niche yet.

Let us notice that the cause of the change of the colour of the vase is not the action of putting or repainting it. This action is merely a collection reinforced in a given situation as a result of the developed sculpture. The cause would be a strong desire of man expressed as reinforcing the collections corresponding to this desire.

Example with elementary particles

Typical elementary particles are collections of very little information concentration and of a strong sculpture. However, these features can refer not only to particles.

Similarly as in the previous example, I have described the situation observed by man (a high level of interpretation) with italics. Whereas I have described the Secral's interpretation of the situation (a low level of interpretation) with normal type.

A particle is moving, we observe the trajectory of its flight.
People see the registered picture of the situation with some delay. From their point of view, they do not immediately take part in the event.

We can observe only selected points from the trajectory of the particle flight. These points represent the successive strong acts of identity. A group of people are observers of the situation. The situation, which each of them observes is similar, since the people participate in the common time sequence. What one can say about the time sequences dealing with this situation has been mentioned in item one of the former example.

Let us consider one point of the particle flight.

1) We observe a collection, which we interpret as a particle. The collection exchanges information with others. Different collections occurring in the area of the information activity of the particle are reinforced. The best separated part of this area is the particle potentials area.
2) One of these reinforced collections will be interpreted as the next element of the time sequence, i.e. the next stage of the particle motion.

When we observe the particle 'flight', the operations of identity are so weak (the quantity of information being transferred is insignificant) that the successive collections being included in the time sequence differ only in some characteristics of the interpretation of canvas. We interpret the situation as the observation of a particle of constant features such as rest mass moving in the space. In every point of its 'flight' we identify the particle by means of a collection of the very same individual sculpture. This creates an impression as if it were still the very same particle. It is difficult to talk here about the motion of the particle in our everyday understanding. Different collections are interpreted as the only and the very same one although finding itself in another place and in another physical time. On the other hand, let us remember that although the statement 'different collections' is a more basic interpretation than that of the time sequence but it does not belong to the lowest level of interpretation.

A particle approaches another one and collides with it.
    The canvas of particles and of the other collections exerting an influence on the situation is interpreted as the observation of two particles whose reciprocal distance is decreasing. Their position, velocity and energy result from the further interpretations of canvas. The more intense informa-

tion interchange between particles brings about the violation of symmetry of their potentials areas. As a result of this, we observe physical acceleration of both particles. We regard this moment as the moment of collision. In addition to acceleration we will most probably observe considerable changes of directions and energy of the moving particles.

From the point of their collision, three particles different from the previous ones are moving away. We interpret that in consequence of the collision of two particles three quite different ones have been produced. Two particles, which collided, have ceased to exist.

As a result of the intense information interchange, the potentials areas of particles have been intensely violated in the course of the approach of particles. The potentials area is a set of collections, which can be included in the time sequence. A stronger violation of this area, i.e. reinforcing some of its components resulted in including collections of the individual sculpture different from the previous one in the time sequence. This has been interpreted as disappearance of one particle and appearance of another one. The change of a number of particles is a result of reinforcing one collection from the potentials area and its inclusion in the time sequence of the niche. Thus, this is not the 'creation' of quite a new particle but reinforcing the existing one and including it in the time sequence of the niche. Speaking here about the time sequence of the niche I mean not only the sequence of the niche of the observed situation but also the individual time sequence of the man's niche as well as every other sequence in which the mentioned particles appear.

Let us notice that <u>not every niche must interpret the same situation in the very same way</u>. If in our time sequence we observe three particles then in another time sequence participating in the situation we may observe another number of particles and even of a different sculpture. This depends on the intensity of the information interchange between these niches and particular collections and on the sculpture of particular niches. Considerable divergences of the interpretation of the very same situation mean a big difference between sculptures of the niches and a feeble connection of both niches.

From the people's point of view, the situation of the particle collision occurred for a very short time and man did not participate in it. From the Secral's viewpoint the situation was the culminating point of the previous reinforcement of the definite collections. This was the culminating point since it was accompanied by the increase in the intensity of the information interchange. The particle collision itself as an event did not differ qualitatively from its ordinary 'flight'. The last one was founded on successive including a collection of a similar individual sculpture (e.g. of the sculpture of an electron) in the time sequence whereas the collision relied on including representatives of other sculptures. Moreover, not in every time sequence the situation was interpreted as the change of the particle sculpture.

It is also difficult to indicate distinct time limits of the event since reinforcing the selected collections from the potentials area preceded it. In this manner, the event occurred earlier but with a less intensity of information. Other collections than the observed ones at the end of the event could participate in reinforcing the potentials areas. Also people could take part in it. If they reinforced similar situations performing operations in their individual time sequences (in their imagination) they reinforced sculpture for such a kind of situation. Thus, they could exert a direct influence on the observed interactions between particles, not by exerting any force on particles but by forming a sculpture for the observed situation.

Self-Creating Language

Example with an electron, a proton and an atom

From the Secral's point of view, the situation is very similar to those described in the previous examples. However, we will try to pay attention to some additional aspects.

Similarly as in the previous examples I have described the situation observed by a man (the high level interpretation) with italics. Normal type is used in the Secral's interpretation of the situation (the low level of interpretation).

An electron and a proton move independently of each other, we observe the trajectory of their flight, they approach each other.
> The interpretation of their motion is similar to the previous example.
>
> Particles more intense exchange the information with one another and this violates the symmetry of their potentials areas.
>
> The selected collections from the potentials area are reinforced during the information interchange. If the situation which we observe begins to resemble one of those contained in the time threads of the sculpture of the niche, it is most probable that a representative of this thread will be included as its element in our time sequence. In this manner, the observed situation will assimilate to one of the well-known in our niche. In this case, a collection representing an atom of hydrogen will be the most reinforced collection from the potentials area. If the influence of the sculpture is significant, the individual collection of the atom will be one of the collections represented by the sculpture of the atom developed within the sculpture of the observers' niche and it will not be exactly the same as that reinforced.

An electron and a proton collide. They unite forming an atom of hydrogen.
> A collection representing an atom of hydrogen has become

so strong that it has been included as the next element in the time sequence of the observed situation. The information interchange between the niche and both a proton and an electron weakens; therefore the particles are not included in its time sequence. Their information interchange with other niches can be continued even on a high level but they do not already belong to the time sequence of the observer's niche. From his point of view, the particles have ceased to exist. Similarly one can say about an atom of hydrogen which although existed in other niches before, from the observer's point of view, was 'formed' when the particles collided.

The observed situation have been forced by the sculpture of the observer's niche. An observer belonging to another niche could find particles of another kind and of another quantity.

We observe the existence of an atom of hydrogen interacting with other particles. We think that an atom is an electron circling round a proton. We can satisfactorily describe it with the equations of quantum physics.

The information interchange between the collection of an atom of hydrogen and its surroundings (a superior niche) remains on a constant level. Collections of the very same individual sculpture of an atom of hydrogen are included in the time sequence of the observer.

An atom can be interpreted as a whole or a composition of other collections. In order to interpret it as a composition, the observer would have to intensely exchange the information with one of its component collections. This would bring about reinforcing the component collection and disturbing the potentials area of the atom. In this manner, too energetic attempt to separate a part of the component collections can result in the change of the individual sculpture of the collection and then it would cease to be perceived as an atom. Less energetic attempts can result in a momentary separation of other collections within the collection of an atom. If the separation is a situation represented in the sculpture of the niche, the latter will decide which component collections will be separated. The way in which an atom of hydrogen 'originated' is of no importance since, as we have explained before, its origin did not consist in connecting a proton and an electron but in including an already-existing collection of an atom in the time sequence of the observer's niche. The collections which will be separated as component ones of an atom are not the very same which participated in the situation in which an atom had appeared. The separated component collections can have quite a different sculpture than a proton and an electron. The sculpture the observer's niche decides which particles will be separated within an atom. If we apply a non-typical method which has not been consolidated in the sculpture to separate particles we will be able to observe quite different collections than an electron and a proton within an atom. The classical description of an atom structure is applied to typical situations consolidated as the threads of time in the sculpture of our niche.

Example with society

I give this example in order to pay attention to the fact that everything what we observe is collections, namely, also we ourselves, people, society and anything we do. Although on account of our emotional connection with these concepts they may seem to be something exceptional to us but they are also subject to the very same laws as the other collections.

Some day one man begins to think about organizing a ceremony.
If this thought about a ceremony is a conscious thought this

denotes that it has been included in the individual time sequence of the man. Other people do not have an identifier of this ceremony yet or they have it in a weak form. For this reason, the collection is not included in their time sequences or in the time sequence of the niche common to people. The thought about the ceremony is not known to other people.

As the thought is being reinforced, its identifier is sent to other collections both internal and external ones for the man. All the collection with which the given man exchanges information can take part in it. The identifier can be sent to other collections during the information interchange irrelevant to the ceremony. If the sent identifier is not too extended, the collection which has received it does not have to be aware of it, i.e. the information does not necessarily have to be included in its time sequence. However, if one of the collections already has a similar identifier, this can be identified with the identifier of the ceremony. As a result of this and of unconscious reinforcing, the identifier of the ceremony can appear in the time sequence of another man. From the man's point of view, it will be his own thought about the ceremony. If that man is able to analyse this thought more accurately, he will read the identifier of the man who reinforced the thought before as one of its well extended component collections.

A man talks about his idea to other people.

A man continues to reinforce the collection representing his thought about the ceremony. He sends the identifier of his thought in a more intense way to the selected people. The very act of the information interchange, if intense enough, will be included in the time sequences of the people participating in the exchange and in the common time sequence of their niche. Its inclusion in the common time sequence most often means a way of the adoption of a secondary interpretation of this phenomenon. The way is forced by the sculpture of the niche (if the information concentration in the sculpture of the niche is greater than in the collections included in the time sequence). In case of people, the sculpture in which, among others, a picture of our body has been developed will force one of the means of communications, i.e., the situation of a conversation, sending visual signals, etc. Thus, if a man wants to transfer information more intensely to another one, a collection illustrating their physical contact, i.e. the interpretation founded on the canvas will be included in the time sequence of their niche. A man desiring to exchange information with another one intensely will begin to reinforce the selected collections from his potentials area and which illustrate such transferring, and on account of the sculpture of the niche he will be able to choose only a limited number of possibilities, e.g., a conversation or writing. Let us notice that the cause of the information interchange is not a conversation. A picture of the conversation is a secondary interpretation of the act of the information interchange.

A group of people prepares and celebrates the mentioned ceremony for the first time.

A collection representing the ceremony has been reinforced by people and in a consequence of this, it has been included in their common time sequence and in each of their individual sequences. Before this happened, the collection of the ceremony had existed in the individual time sequences of people. The fact that the collection could begin to exist in time sequences was preceded by appearing this collection in the potentials areas of people and in their common niche. Observing the potentials areas one could find in advance that the ceremony would take place and one could also determine its course.

Self-Creating Language

The ceremony is repeated in the same season every year. It is repeated for many years.

The repetition of the ceremony contributes to the development of the thread of time depicting the situation in the sculpture of the niche. The thread connects the definite season and place with the actions performed before, during and after the ceremony.

If many people took part in the ceremony and if it was repeated many times, one could expect that the thread associated with the ceremony had been strongly developed.

A group of people determines to stop celebrating the ceremony. This can be given in the form of persuasion, official ban or other.

A new collection is being formed (as a composition of other collections existing before). The interpretation of this collection will depend on the fact how the opponents of the ceremony imagine it. The collection may represent a picture of society that does not celebrate the ceremony. It can also represent a picture of canceling the ceremony. As newly-arisen the collection is feebly extended. It is strong enough to exist in the individual time sequences of some people but it is not strong enough to appear in their common time sequence. Thus, meanwhile the collection is observed as a thought.

Whereas, the collection representing a ban (verbal or written one) differs from the mentioned above. It is the interpretation of the information interchange between people on the level of canvas.

In spite of attempts to stop it, the ceremony is still celebrated. One thinks that this is a contribution of its present advocates.

The collection of the ceremony is so strong that it is again included in the common time sequence.

Let us notice that it would be difficult to prevent the appearance of the collection in the time sequence even if the participants of the ceremony tried to do so themselves. Then other people would take part in the ceremony. At the moment it is not important who is the advocate and who is the opponent. However, it is important for forming the time sequence in the distant future since collections are not able to significantly change the potentials area of the common niche in a short time. As long as they do not do it, the thread of time connecting the circumstances of the ceremony will bring about its appearance in the time sequence. The collections formed by the advocates and opponents of the ceremony are reinforced by every man who thought about them at least for a moment. It is not important whether the given man regards himself as an advocate or not but it is important which collections he reinforces, i.e., about what he thinks.

The situations, at the moment when they are observed, manifest the collections from the potentials area earlier reinforced. At the moment they 'are coming to existence' (observed by people) they do not bring about events directly following them. As the cause of events one can regard much earlier reinforcement of proper collections from the potentials areas. The cause of any event had arisen and reinforced much earlier than the inclusion of an event in the time sequence of the common niche, i.e. 'it appeared in reality'.

The celebration of the ceremony weakens after a long time. In spite of attempts made to return to regular celebrating the ceremony, the celebrations are stopped but they are remembered.

The earlier ceasing to reinforce the collection of the ceremony may be the cause of ceasing celebrations. Every repeated ceremony results in the information interchange between it and other collections. In the situation when the collection is no longer supplied it contains less and less unique information possible for transfer. Its chance to appear again in the time sequence of the niche grows less, since its information potential with regard to the niche decreases.

Another cause of ceasing celebrations may be the collection reinforced by the opponents of the ceremony. The collection formed by them could adopt a different form according to their not-always-conscious intentions. If they concentrate on forming the picture of their victory and on destroying the ceremony, the collection representing this situation will appear in the time sequence, since what the collection represents is a short-lived event and so it will be observed. However, if the opponents concentrate on the picture in which a society does not permanently celebrate ceremonies, the collection after its inclusion in the time sequence will be also observed in such a manner. Its reinforcement determines how often it will be included in the time sequence, i.e., how long it will be observed by people. In this period of time, the collection of ceremony as feebler one will not be included in the time sequence. A strong collection of the opponents of the ceremony can exclude the collection of the advocates from the time sequence for some time but will not bring about weakening the collection of the ceremony. In case, when both collections are strong, two separate time sequences corresponding to two niches of the advocates and the opponents can be isolated. The advocates of the ceremony will celebrate in their time sequence whereas, at the same time, the opponents will experience their lives without ceremony. In the interpretation of people this will be observed as the territorial isolation of both groups living according to their principles. Such a situation will be possible if the remaining part of the society remains passive towards clashing the desires of two groups. The society is a superior niche for these groups. What scenario of the events will be observed depends on the fact how developed the potentials areas have been by the earlier actions of all collections. From the point of view of the low levels of interpretation, <u>the events will occur so that the maximum equalization of information potentials between collections could be possible</u>. Whereas, from the point of view of the high levels of interpretation, we can say <u>the events will occur so that all the desires of every collection would be fulfilled, at least in the individual time sequence if not in the common one</u>. What we understand as desires and their fulfilment has been described in the chapter 'Desires, emotions'.

After some time ceremonies are renewed and they are again regularly celebrated.

If ceasing to reinforce the collection of the ceremony was the reason for ceasing the celebrations it means that reinforcement was renewed and lasted long enough to give facilities to cause the appearance of the ceremony in the time sequence of the niche.

If the inclusion of a stronger collection in the time sequence of the niche was the reason for ceasing to celebrate the ceremony this denotes that the information potential of that collection has weakened.

In general, the described mechanism deals not only with ceremonies but also with any actions, any collections and not only with people. The example illustrates the effect of forming the sculpture of the niche.

Comparison of the Secral with computer science programming languages

We are accustomed to the fact that a programming language is a set of definite commands and data arranged according to a definite syntax.

In the Secral as a language we can distinguish only one

command: *identity*. If one sought for its contrary, it would not be the operation of differentiation since it is only the interpretation of the operation of identity (I have described it more closely in the chapter 'Discrimination'). As an opposed form one can regard merely Area B since it is not a collection. However, it is not the contrary since the concepts 'Area B' and 'the contrary' do not reach either the real Area B or the low levels of interpretation. The operation of identity does not have its contrary.

It is difficult to call the Secral a programming language. It cannot be fully implemented as a computer program. Programming languages and computers are interpretations formed on the high levels whereas the Secral covers also the low levels of interpretation. A description of the Secral, i.e. a description of the present kind but not as a whole can be implemented.

However, it is possible to compare the features of the Secral to those of the known programming languages. Maybe it will be helpful to understand some Secral's ideas or it will help people to construct programming languages.

The action of the Secral resembles a little the principles met in modern object programming languages. Let us mention some analogies and differences:

1) Encapsulation.
   It is an idea resembling the idea of niches. The difference between them is that the niche limits are dynamic and their range varies. Their clear differentiation can also change.

2) Classes, methods, objects.
   A class is the equivalent of a niche to a certain extent. All the methods contained in a class are the sculpture of the niche. An object is an individual collection.
   If we desired to bring the object-oriented language closer

to the Secral, the division into the three concepts should not be rigid. In some situations, an object can play a part of a method or a class. The class can be interpreted as a method or an object and inversely.

In the Secral, the action can be the object of the action, whereas the object can be treated as the action. This depends on the way of interpretation adopted by the given niche.

3) Inheritance.

It is similar to the role played by hierarchies of the niches including their sculptures. The visible difference is the fact that the sculpture of the niche is being developed together with every operation performed in the niche. The way the hierarchy comes into being is also different. The hierarchy of niches is not imposed beforehand but it is shaped according to the differences in information concentration of collections and to the intensity of information exchange between them.

4) Principles of language structure, syntax.

In the classical programming language they are imposed in advance. They are written in a compiler or in an interpreter for good. If we desired to draw such a language nearer to the Secral, the semantics should be formed like the sculpture of the niche, i.e. during a long-lasting action of the program in real circumstances.

5) Data.

As it has been mentioned above, the Secral does not distinguish precisely between an action and an object. Data can be the functions of an active program. The elements of the program can be treated as data.

Knowledge

In everyday meaning knowledge means that one has all the encyclopaedic information on the given object and that one is provided with a theoretical description (model) of the object. From the Secral's point of view, the mentioned above way of understanding knowledge is proper only on the high levels of interpretation and only in Area K.

Assuming that we are talking merely about knowledge dealing with Area K, by knowledge we will understand the conscious action of a collection founded on understanding its own situation.

We can distinguish four basic kinds of knowledge:

1) One has the system of knowledge treated as a set of concepts. Such a system constitutes a model of reality. Its characteristic feature as a collection is that it is less extended than the collection, which is its owner. Owing to this, the system can be easily transferred between collections and in our case, between people.

   This kind of knowledge is represented by science, religion and all the other systems spread between people through the information transfer founded on the interpretation of canvas, i.e. by means of words, writing, pictures etc.

2) One has conscious access to information contained in one's collection, i.e. the collection with which the observer identifies himself. The collection with which an individual man identifies himself is much more extended than any system of knowledge exchanged between people, even this part of the collection, which is connected with our physical mind of which we are aware every day. The collection of man contains information gathered during the whole history of his existence and which did not begin together with his appearance in our niche but much earlier. In everyday life, the man identifying himself

strongly with his physical mind has limited access to the whole of knowledge. From such a point of view, it occurs in the areas, which we call deep subconsciousness and in even deeper areas. As an extended collection, man is able to change the interpretation to which he is subject. He can acquire access to any information both within his collection and within the whole of Area K. This kind of knowledge is intermediate between that described in the previous item and the next one. It differs from an ordinary system of knowledge in the fact that it constitutes knowledge of an individual collection and that it is not subject to firm generalizations.

3) One identifies oneself with the collection representing the object of one's interest. Knowledge is being acquired at the moment when it is needed and there is no need for appealing to one's memory. This kind of knowledge is not based on the interpretation of time.
In this case, the intermediate models of phenomena are omitted and every subject of knowledge is treated individually without any generalizations.
Knowledge of this kind cannot be transferred to another collection in the form in which it is known. Every transfer will consist in sending an identifier of the event which will be merely the interpretation of an original situation but not full information about it. Transfer of knowledge of this kind will change it into one of the kinds described in the previous items.

4) One resigns from any interpretation or one comes to know all the possible interpretations (the whole of Area K). The knowledge is connected with recognition of Area B. The knowledge cannot directly be transferred to the collection and it cannot be described but the way of acquiring the knowledge can be transferred.

The succession in which the kinds of knowledge have been mentioned is not accidental. We began our considerations from the knowledge occurring on the high levels of interpretation and we gradually descended to the lowest ones. The lower the level of interpretation is, the less information we lose making interpretations. Thereby knowledge assumes a more and more effective form.

Ways of acquiring knowledge, systems of knowledge

Let us consider the descriptive kind of knowledge (item one in the previous chapter) since we can impart this kind of knowledge to each other.

From the Secral's point of view, the essential factor of knowledge is the level of interpretation with which knowledge deals and on which level of interpretation the collections which have knowledge occur. Plainly speaking, it is important to have accurate concepts, to understand them and to apply them not only to the theory but also to perceive the world taken as a whole and to act in it.

As we know from the previous chapters, logic which we use (concluding, proving rightness or injustice) is not the only way of thinking and describing phenomena. Logic finds itself on too high level of interpretation to serve as a criterion for evaluating the system of knowledge. This does not denote that we should reject it but that we ought to realize what it is and where its limits are. The systems of knowledge are not only systems founded on logic and not only those based on the so-called empirical experience. <u>A system of knowledge is every separated set of concepts</u>.

The existence of many systems of knowledge can incline one to devise the way of evaluating these systems. In science one ordinarily assumes 'its accordance with the empirical experience' as a criterion for evaluating the system of knowledge. A personal conviction can serve for the criterion of such evaluation in the systems of knowledge, which are founded on logic to a lesser degree.

In the Secral, both a personal conviction and empirical experience are two cases of the very same situation but they differ only in the intensity of the information transfer. The empirical experience is such information transfer in which the sculpture of the common niche plays a decisive part. The sculpture of the niche is a reinforced 'personal conviction' of many collections strong enough to form the mentioned sculpture. On the other hand, an individual collection is able to perform an act of identity covering any niche and even the whole of Area K. If such an act enables a definite 'personal conviction' to have, this conviction will be founded on knowledge which can be more correct than the empirical experience known within a certain niche. The empirical experience is the observation of the common time sequence and it constitutes the interpretation ignoring a part of the accessible information. Such an experience gives facilities solely for obtaining a picture of the present sculpture of the niche with which it deals.

Treating the empirical experience as a means for estimating the system of knowledge which better, i.e. more correctly describes phenomena than others, is not always right.

From the Secral's point of view, the best criterion for evaluating the system of knowledge is its level of interpretation. The lower the level is, the more precisely the structure of concepts is defined. The system of knowledge, formed on the level of interpretation higher than the first one, as a collection incessantly varying, is not able to describe phenomena correctly since it is only their definite interpretation. The more noticeable it is, the higher level of interpretation is on which the system of knowledge has been formed.

A means of producing a more correct description is simultaneously accepting many different systems of knowledge to a degree corresponding to the level of interpretation of the given system. If it is difficult to accept them simultaneously, we will obtain a similar effect by learning how to freely move between different systems of knowledge and not treating the selected one as the only 'true'. One resigns from knowledge when one places any systems of knowledge opposite to each other as competitive or excluding ones. Even a system regarded as the most improbable one is a product of the collection from the given niche and for that reason it includes some elements of the correct description of the situation. A criterion, which enables the usefulness of the system to evaluate, is its level of interpretation, i.e. the way in which the concepts of the system are developed. However, if we are interested only in a correct description of the given niche, we should adopt the agreement of the system of knowledge with the sculpture of the niche, i.e., what we call 'the empirical experience'. Such a system can be successfully utilized, first of all, in the niche for which it has been formed but only for some time until the sculpture of the niche varies significantly. One should remember that a change in the sculpture of the niche, although slight, occurs constantly together with every act of identity which concerns the niche. A multiple act of identity with a definite system of knowledge can bring about such a change in the sculpture of the niche that the given system of knowledge will be perceived as one satisfying the conditions of the empirical experience. Such a case has been described in the chapter ' Particle physics as niche'. The causative role of theorems, described in the chapter 'Theorem, truth, untruth' also deals with the systems of knowledge.

How should the Secral language be understood?

There are two methods permitting the laws of collections to understand:

1) One rejects all the secondary concepts, i.e. one comes to

know a collection on the first level of interpretation.
Such recognition is full but it cannot be imparted to another collection since it requires making use of concepts, i.e. one's passage to the high level of interpretation.

2) One attempts to describe the laws by means of the high level of interpretation.
This way is used in the present book. It allows a description to be imparted to another collection. Such a description will never be perfect since we use the local concepts developed in our niche. However, we should try to make it the best. As a general principle, we can regard that while speaking about a collection we must not forget all its properties known to us. While describing a situation, I usually paid attention to the most important feature of a collection at a given moment still bearing in mind its other properties. We have not studied all of them all the time in order not to make a vague description. I ask the reader not to forget its other features of the situation, also those, which seem to be less important. Without them it is easy to produce a unilateral description. <u>At the same time, bearing in mind all the features of the collection known to us and not separating them from each other will compensate us for the disadvantage of staying on the high levels of interpretation to the greatest degree.</u>

Let us remember that logic and understanding founded on logic have originated on high levels of interpretation and they may be useful if we want to impart our knowledge to someone else. However, one's attempt to understand the first level of interpretation fully with use of logic and to understand the whole of Area K will fail. This will be so not because logic is defective but because it is a product of the high level of interpretation. As such, logic is the result of a loss of some information and in consequence of this, it constantly varies.

Self-Creating Language

We are not helpless in our understanding a collection since logic is not the only means of reasoning. As collections we have a means of reasoning independent of the levels of interpretation, this is identity. We may also fully understand collections performing an act of identity with the whole of Area K or restraining ourselves from any interpretations.

The Secral as a system of knowledge may seem to be reductionism, i.e. explaining different phenomena by means of simple laws. It is only partly true. It is not true, because from the Secral's point of view, a single system of knowledge cannot describe accurately any phenomena. Here the Secral is not an exception. On the other hand, it may be treated as some instructions for showing how an individual collection (conscious entity) can practically understand the situation in which it occurs.

One should not treat Secral as a representative of some '-ism' such as conventionalism, empirism, etc., since it accepts all the ways of thinking as expressions of definite interpretations.

What results from the presented description?

The Secral permits phenomena to be described in an avalanche way following the interpretations of the simplest concept which is a collection increasing analogously in an avalanche way. It includes the refined wealth of the world which we experience and even more. I am not able to describe all the interpretations and, at the same time, I should not like to deprive the reader of this pleasure so I will stop here.

In the book I have avoided describing phenomena by means of mathematics. Even in the simplest form of arithmetic, mathematics is a secondary interpretation of a collection. We as people are also collections. To understand this and laws governing collections it is enough for us to make use of these laws according to our needs and will. A mathematical description would permit the persons who do not understand it to take advantage of the knowledge to a certain degree. Maybe this action would be right but I do not regard it as a superior objective. As a superior objective I regard making independent understanding easier for everyone.

The most general conclusion which may be drawn after learning these laws, is the unrestricted action of every individual man. There is a possibility of realizing every thought, since it does not qualitatively differ from realization itself. They differ only in their intensity. We may change the sculpture of the niche, i.e. the laws of nature and physics. We have the opportunity to learn and shape the future. We can do all this if we find a reason for this.

Understanding the laws of a collection gives facilities for making use of the knowledge in practice. Moving to more basic levels of interpretation it is not necessarily for us to build machines and devices, which might do something on our behalf. The devices are useful only within the framework of the interpretation of canvas. We, being collections, do not have to create next collections in order to obtain still other ones. Other collections are not brakes for our actions. Our submission to interpretations is the only brake. We have access to all the collections from Area K in a more or less intermediate way. Keeping oneself busy at collections and forming new ones means that one is permanently moving in the very same environment. This seems to be sensible until we divide collections into better or worse ones, worth obtaining or not. Differentiation of this kind, like every interpretation results from a loss of a portion of information. If we do not reject the information we will come to the conclusion that every collection is simply a collection with no particular features. The features resulted from our previous ignorance.

Resignation from any interpretation is the only move, which will not bring about further involvement in overlapping successive interpretations. Thereby we take a step towards Area B. This is the most priceless opportunity, which we have since it is the only one qualitatively different from keeping oneself constantly busy at collections.

CONCLUSION

To end with. I should like to warn both myself and the reader who will want to interpret everything what I have described in the book not to jump to conclusions.

We remember the theory of atoms formed in ancient Greece. Thousands of years later our science discovered molecules, which to an illusion resembled those described by the ancient Greeks as the smallest indivisible elements of matter. With joy of their discovery they were called atoms - just like the ancient Greeks called them. Before long we discovered that atoms are neither indivisible nor the smallest particles. In this case, these were not the ancient Greeks who were mistaken but we who in a hurry interpreted our discovery.

While writing this book I have tried not to make a similar mistake and I ask the reader for the same.

The author

Self-Creating Language

More important concepts used in the book

CONCEPT, PAGES

Ambiguity of identifiers, 27
Area B, 49
Area K, 49
Area of concentration, 42
Basic mind, 100
Basin of collection, 29
Canvas, 71
Collection, 21
Connection between collections, 56
Countability, 25
Extended collection, 30
Extended identifier, 30
Hierarchy of niches, 25
Identifier of collection, 14
Individuality, individual collection, 25, 33
Information concentration, force of collection, 30
Information potential, 28
Interpretation, 19, 69
Level of interpretation, 19
Niche 23
Physical mind, 99
Point of time, 41
Potentials area 36
Principle 1, 9
Principle 2, 9
Repeatability, 43
Sculpture, 31
Similarity, 26
Situation (state of niche), 41
Supercollection, 134

Supplying collection, 30
Thread of time, 42
Time key, 43
Time sequence, 36
Timelessness, 40

Synonyms of concepts

The concepts which are used interchangeably in the book when it is necessary to emphasize one of their features:

1) Collection = act of identity = information = information transfer.
2) Interpretation = concept = loss of information.
3) Collection force = information concentration.

Index

A

Absolute   59, 71
Acceleration   71, 72, 73, 74, 75, 76, 77, 79, 148. *See also* Conservation of acceleration
Act of identity   17, 18, 30, 31, 32, 38, 39, 40, 41, 43, 47, 48, 52, 57, 61, 64, 65, 66, 67, 77, 80, 83, 89, 92, 93, 97, 98, 99, 102, 103, 104, 105, 109, 110, 120, 125, 130, 132, 134, 138, 146, 164, 165, 167
Ambiguity of identifiers   27, 31, 33, 34, 39, 43, 81, 83, 96, 128
Ancient Greeks   170
Animals   134, 143
Archetype of spaces   26
Area B   49, 50, 51, 59, 68, 103, 139, 140, 159, 162, 169
Area K   49, 50, 51, 59, 62, 63, 66, 68, 79, 82, 85, 97, 98, 103, 113, 139, 161, 162, 164, 166, 167, 168
Area of concentration   42
Arithmetic   26, 85, 167
Artificial intelligence   109, 110
Atom   150, 151, 152, 170
Averaging   73, 74, 75, 79
Axioms   85, 95

B

Basic mind   100, 106
Basin   29, 30, 31, 34, 70, 90
Beginning of existence   45, 46, 61
Behaviour of people   101, 108. *See also* Critical situations; Habits
Bit   10, 64, 65
Body   70, 72, 88, 98, 99, 100, 102, 103, 105, 106, 110, 111, 122, 154
Brain   102, 122

C

Canvas   71, 72, 73, 74, 75, 76, 77, 78, 79, 93, 97, 98, 99, 100, 105, 106, 109, 110, 111, 113, 122, 125, 126, 129, 132, 133, 135, 138, 147, 154, 155, 161, 168
Causative role of theorems   58, 165

Cause   50, 59, 61, 63, 64, 138, 141, 146, 154, 156, 157
Cause and effect   59, 60, 141, 142
Ceremony   152, 153, 154, 155, 156, 157, 158
Character of man   111, 128, 130
Chemical compounds   134
Collection   13, 14, 15, 16, 18, 21, 22, 23, 24, 25, 27, 28, 29, 30, 31, 33, 34, 36, 42, 44, 45, 49, 61, 64, 67, 78, 93, 105, 114, 116, 125, 129, 164, 166, 167, 168
Communities   107
Complexity   23, 35, 40, 62, 94
Computer   109, 158, 159
Computer disk   109
Computer processor   109, 110
Computer science   64, 158
Concluding   85, 86, 163
Connection between collections   56, 57
Consciousness   18, 61, 67, 70, 80, 89, 98, 99, 101, 103, 107, 108, 110, 121, 131, 134, 138
Conservation of acceleration   79
Conservation of energy   79
Continuum   26, 45, 65, 76
Contrary of the good   126
Conventionalism   167
Conversation   90, 154
Countability   25, 26, 35, 43, 45, 67, 70, 77, 94, 95, 97
Criterion for evaluating the system of knowledge   163, 164
Criterion of correctness   85
Critical situations   137

D

Data   158, 160
Desire   116, 117, 118, 119, 120, 121, 122, 123, 124, 125, 126, 129, 130, 143, 145, 146, 158. *See also* Fulfilment
Destiny   140, 142
Dimensions   26, 76, 77, 78, 80. *See also* Three dimensions
Discrimination   12, 14, 26, 34, 36, 70
Distance   36, 65, 71, 72, 73, 75, 76, 77, 78, 127, 147
Dogma   86
Dreams   54, 55, 111, 113

E

Effect   27, 42, 58, 59, 121, 129, 141
Ego   123
Electromagnetism   75
Electron   91, 149, 150, 151, 152
Elementary particles   76, 150
Emotion   76, 101, 111, 116, 117, 118, 119, 120, 121, 122, 125, 126, 127
Empirical experience   163, 164, 165
Empirism   167
Encapsulation   159
End of existence   45, 46, 48
Energy   73, 74, 75, 78, 79, 147, 148. *See also* Conservation of energy
Entropy   66, 76, 108
Equalization of information potentials   158
Euclidean axioms   95
Everyday interpretation   20, 38, 39, 45, 46, 53, 57, 59, 69, 75, 99, 116, 126, 127, 161
Example   143, 146, 150, 152
Example of expressing desires   129
Example with a man and a vase   143
Example with an electron, a proton and an atom   150
Example with elementary particles   146
Example with society   152
Extended collection   30, 51, 99, 100, 110, 114, 121, 122, 142
Extended identifier   30, 139

F

Finger-marks   132, 133
Flowers   128
Force of collection   30
Fortuity   63, 64
Free will   138, 139, 140
Frequency   38, 41, 74, 75
Fulfilment   123, 124, 158

G

Genetic code   106, 111
Geometry   85
Glass   60
Global point of view   139
Gravitation   75
Greece   170

H

Habits   131
Hierarchy of niches   25, 39, 42, 51, 110, 131, 160
Hypothesis   58

I

Identifier   11, 12, 13, 14, 15, 22, 27. *See also* Ambiguity of identifiers
Identity   12, 17, 18, 19, 26, 27, 28, 31, 36, 39, 40, 67, 159
Imagination   54, 56, 81, 83, 113, 114, 115, 116, 118, 119, 124, 128, 149
Imagining in two ways   115
Individual existence   67, 134
Individuality   25, 33, 34, 86, 92, 119, 122
Information   10, 11, 13, 15, 18, 22, 28, 30, 64, 65, 109, 134
Information concentration   30, 49, 67, 87, 88, 89, 92, 134, 135, 143, 146, 154, 160
Information negligence   13
Information potential   28, 36, 37, 44, 45, 57, 76, 79, 89, 90, 112, 115, 116, 118, 119, 127, 140, 142, 157, 158
Information transfer   16, 17, 19, 22, 28, 65, 66, 76, 90, 107, 161, 164
Inheritance   160
Internal and external collections   15
Interpretation   13, 18, 19, 21, 25, 35, 40, 43, 49, 50, 54, 67, 69, 70, 71, 167, 168

J

Joy   100, 117, 120, 170

Judgement 126, 127

K

Kinetic energy 79

L

Laboratory 9, 10, 11, 12, 13, 15, 16, 18, 25, 46, 67, 69
Laws of nature 83, 84, 85, 96, 97, 129, 168
Level of interpretation 19, 20, 70, 71, 166
Logic 17, 48, 83, 85, 86, 87, 96, 135, 163, 164, 166, 167
Loss of information 19, 21, 60, 78, 81, 96

M

Macroscopic objects 76
Making gestures 132
Man 81, 83, 86, 88, 89, 98, 99, 100, 101, 102, 103, 105, 106, 107, 108, 110, 112, 114, 117, 118, 121, 122, 123, 130, 134, 137, 138, 139, 143
Mankind 91, 111, 112
Mathematical constants 83, 85, 95, 96
Mathematics 57, 85, 95, 167
Matrix 32, 33, 111, 131
Matter 93, 94, 95, 170
Means of communication 136, 154
Measurement 38, 39, 41, 46, 71, 72, 98
Memory 51, 52, 88, 99, 112, 162
Minerals 138
Morality 125
Motion 73, 74, 79, 147, 150
Mould 32
Multidimensional 73, 74, 78

N

Naivety 16, 19
Negation 129, 130
Niche 23, 24, 25, 34, 40, 51, 87, 110
Nuclear forces 75
Number (concept of) 26

Number pi   85

O

Object-oriented language   159
Obviousness   20
Operation contrary to identity   17

P

Painter   128
Paradox   13, 81, 83
Particle   76, 78, 87, 89, 90, 91, 92, 93, 134, 146
Particle flight   146, 147
People   20, 21, 57, 78, 86, 87, 88, 90, 94, 99, 101, 102, 103, 107, 108, 109, 110, 111, 112, 113, 114, 118, 120, 121, 125, 128, 130, 131, 132, 133, 134, 135, 136, 137, 138, 142, 143, 146, 149, 153, 154, 155, 156, 157, 158, 161, 167
Personal conviction   163, 164
Physical constants   83, 84
Physical contact   133, 154
Physical mind   99, 100, 102, 106, 161, 162
Physicists   87, 88, 89, 90, 91
Physics   57, 78, 83, 84, 85, 87, 88, 96, 151, 168
Planet   112, 143
Plants   33, 134, 137, 143
Plum-tree   33
Point of time   41, 65, 66, 70
Potentials area   36, 39, 42, 46, 56, 61, 73, 74, 75, 76, 80, 82, 83, 90, 92, 93, 115, 117, 118, 121, 140, 141, 142, 143, 147, 148, 149, 150, 151, 154, 156. *See also* Symmetry of potentials area
Principle 1   9, 22, 67
Principle 2   9, 25, 67
Principle of conservation   78, 79
Probability   63, 64, 115, 124, 128, 142
Problem of infinity   13
Programming languages   158, 159
Proton   150, 151, 152
Proving theorems   57, 58, 85, 96
Psyche   70, 78, 98, 99, 100, 101, 103, 107, 109, 111. *See*

also Ego

Q

Quantity 20, 25, 26

R

Reality 52, 54, 55, 56, 57, 58, 69, 70, 81, 82, 83, 84, 85, 86, 93, 94, 107, 112, 113, 114, 115, 116, 118, 119, 121, 122, 128, 139, 140, 142, 143, 156, 161
Reductionism 167
Religion 107, 112, 161
Repeatability 43, 44, 55, 70
Research instruments 88
Resigning from knowledge 165
Responsibility 126, 128
Rest mass 73, 74, 75, 79, 93, 147. *See also* Zero rest mass

S

Sculpture 31, 32, 33, 34, 35, 40, 42, 44, 52, 55, 56, 58, 60, 66, 73, 82, 83, 84, 85, 90, 91, 92, 95, 96, 101, 104, 105, 109, 110, 117, 118, 123, 131, 141, 144, 145, 146, 150, 154, 164, 165, 168
Sculpture of collection 33
Sculpture of niche 33
Secral 68, 69, 72, 79, 84, 85, 87, 92, 94, 95, 96, 99, 103, 107, 109, 143, 146, 149, 150, 158, 159, 160, 161, 163, 164, 165, 167
Sense of life 102, 103
Similarity 26, 27, 29, 43, 70, 105, 123
Simple organisms 134
Social groups 107, 125, 128
Society 70, 86, 88, 107, 130, 152, 155, 157
Solitary man 113
Space-time 65, 70, 71, 72, 75, 78, 93, 97, 135
Spaces 26. *See also* Archetype of spaces
Speed 65, 66, 71, 72, 73, 74, 75, 76, 77, 79. *See also* Top speed possible to be observed
Speed of light 79
Subconsciousness 162
Suffering 100, 117, 120

Supercollection 134, 135, 136, 137, 138
Supplying collection 30, 137, 138
Symbol 52, 53, 95
Symmetry of potentials area 148, 150
Synchronizing 78, 113
System of knowledge 70, 87, 161, 162, 163, 164, 165, 167. *See also* Criterion for evaluating the system of knowledge; Resigning from knowledge

T

Talking people 132
The good 125, 126. *See also* Contrary of the good
Theorem 17, 57, 58, 59, 85, 86, 95, 96, 165. *See also* Proving theorems
Thought 54, 89, 90, 100, 101, 102, 114, 116, 118, 133, 135, 136, 137, 139, 152, 153, 155, 168
Thread of time 42, 43, 56, 58, 60, 90, 115, 145, 150, 155, 156
Three dimensions 77, 78
Three-dimensional physical space 77
Time key 43, 54, 130
Time sequence 32, 36, 38, 39, 40, 41, 42, 43, 45, 46, 51, 54, 57, 60, 61, 70, 74, 76, 78, 80, 89, 92, 103, 112, 114, 117, 118, 122, 125, 127, 133, 141, 143, 148, 151, 157, 158
Top speed possible to be observed 65, 73, 74, 79
Travels in time 79, 82
Truth 57, 59

U

Universe 85, 96, 97, 98, 112
Unshakeable 59
Untruth 57, 59

W

Way of thinking 16, 17, 19, 85, 86, 119, 163
Word 17, 52, 53, 54, 57, 59, 102, 129

Z

Zero rest mass 73, 74

ISBN 1553692489